Dragonflight

In search of Britain's dragonflies and damselflies

MARIANNE TAYLOR

BLOOMSBURY

LONDON · NEW DELHI · NEW YORK · SYDNEY

Published 2013 by Bloomsbury Publishing Plc,
50 Bedford Square, London WC1B 3DP

ISBN (print) 978-1-4081-6486-0
ISBN (e-PDF) 978-1-4081-8636-7
ISBN (ePub) 978-1-4081-8637-4

A CIP catalogue record for this book is available from the British Library.

Page design and typesetting by Nimbus Design
Printed in Great Britain by CPI Group Ltd (Croydon), CR0 4YY

10 9 8 7 6 5 4 3 2 1

CONTENTS

Introduction

ৡৼ ৠৼ

What came first, our invention of the dragon, or our discovery of the dragonfly? The concept of a dazzling jewelled creature borne on shining wings, as ferocious as it is beautiful, fits the real insect just as well as it does the mythical beast, if we ignore the small matter of scale. Gram for gram, a dragonfly packs a mighty predatorial punch — a powerful and incredibly agile flyer, with colossal eyes that cannot miss the smallest movement, long barbed legs for snatching prey, and (especially in its infancy) an eating apparatus so bizarre that it has inspired some of our most horrifying fictional monsters.

Dragonflies are considered primitive insects from an evolutionary point of view. Unlike butterflies, bees and beetles, they do not pass through a pupa stage on their way to adulthood, but the adult dragonfly just bursts straight out of the skin of the full-grown larva, or 'nymph' as the aquatic young is sometimes known. When the time comes for adult dragonflies to reproduce, they join together and mate via an extraordinarily circuitous process that is unique in the animal world.

Legends frequently link fire-breathing dragons and distressed human damsels, and so it is with their insect namesakes too. The insect order Odonata contains the dragonflies and a similar number of their small relatives, the damselflies. Like dragonflies, damselflies have long, slim bodies and two pairs of filmy wings, but they are lightly built, fragile-looking, graceful creatures. This delicate appearance belies their nature, for the damselflies are also skilled hunters of smaller insects, although they themselves may fall prey to the bigger dragonflies.

Unlike the dragon of legend, which is born into fire, the dragonfly starts its life in water, and its life as a nymph underwater is usually many times longer than its existence as a flying adult will be. Although a big adult dragonfly has the potential to cover hundreds of miles on the wing, it still needs to find fresh water to breed. Therefore, dragonflies are most often seen hunting around or close to water, whether that is a lake, river, heathland ditch, moorland tarn or garden pond. Damselflies, which fly more weakly, rarely stray far from the edge of the water where they spent their infancy. If you want to see plenty of dragonflies and damselflies, wait until spring and head for a watery landscape.

About 40 species of dragonflies and damselflies regularly breed in the UK. New potential colonists appear from time to time, making it difficult to be more precise about this figure, and there have been extinctions of several Odonata species over the last 50 years. Some of our resident dragons and damsels are common and widespread – go anywhere where there is fresh water at the right time of year and you are likely to see them. Others are scarce and restricted to just one or a few patches of habitat, so a special trip will be needed. To see the vagrant dragonflies that periodically wander here from overseas, you really just need to be very lucky (and be able to get time off work at very short notice).

Though I have been an ardent nature-lover all my life, my passion for dragonflies and damselflies has been a slow-burner. However, two separate experiences stand out as pivotal in my growing appreciation of them.

A FALLEN DRAGON

The summer of 1995 was glorious – week after week of beautiful weather. I remember it better than the others of the 1990s because it was the year I graduated from Sheffield University, and my boyfriend of the time and I spent most of it travelling around the UK in a camper van, looking for butterflies. In between these trips, we stayed with our parents – more often with James's than mine as his mum and dad had a country house, surrounded by dipping and diving Sussex hillsides, and there was no better way to pass a sunny afternoon than out on their patio, contemplating the stunning view while nursing a cold drink.

We had brought two rescued cats with us from Sheffield. Often, our silver tabby would keep company with us on the patio, and do her best to catch the big dragonflies that were themselves out hunting over the garden, chasing down smaller flying insects. She was a young and athletic cat, and made prodigious leaps at the dragons as they rattled past, but didn't really look close to catching one.

I couldn't tell you now which species these dragons were. I knew they were hawkers, or 'aeshnas' according to the rather quaint terminology of my *Reader's Digest* insect book, because they were very big, never seemed to settle, and wore beads of bright green and blue colour in neat symmetrical patterns down their long, parallel-edged black bodies. They had fighter-jet speed coupled with the agility of a hummingbird, and while you could never hope to touch one in flight they would zoom past your ear with millimetres to spare, the dry hard rattle of the fast-flickering wings making you jump out of your skin if you hadn't seen them coming. Although my interest then was in birds and butterflies, it was impossible not to notice and admire these spectacular dragons.

One day, the cat managed a particularly high and well-timed jump, and somehow brought down one of the hawkers, which she immediately tried to subdue with a bite. James rushed forward, grabbed the cat, and eased the struggling insect clear of fangs and claws. I shut the thwarted cat inside and hurried back out, where James was opening his hand to reveal a very sorry-looking dragon.

Bashed and bitten, its wings were crumpled and its dented body had lost its glittering colour. It was still alive though, its skinny stick-legs supporting most of its weight, and it hung on hard to James's hand with six wicked-looking claws. I'd never had the chance to look at a dragonfly close-up before, and I was immediately fascinated. I was used to looking at butterflies at point-blank range, but this was something else. While a butterfly's coating of fluff softens its features and makes it almost mammal-like, almost cuddly, here the details of the creature's structure were laid bare. It seemed assembled from a great toy box of finely machined parts, individually painted and precision-cut to fit together. From its immense wrap-around eyes to the tidy successive sections of its long, slim body, it was extraordinarily beautiful and utterly alien.

As we examined it, we both noticed at the same moment that something strange was going on. Before our eyes the damaged wings were slowly straightening and stiffening, and the squashed part of its abdomen was puffing up to its correct dimensions. Colour was returning to its body and eyes, and it was standing straighter, its abdomen lifting clear of James's fingers. The head suddenly tilted, making us both jump – was the dragon looking back at us? The restored wings began to vibrate, blurring away the exquisite tracery of vein-work, and then the dragon lifted off straight up like a jump jet. It flew powerfully away, hardly a trace of a wobble to betray its recent brush with death.

THE ORIGIN OF BEAUTY

I began to notice dragonflies more after that. However, it wasn't until the following year that their smaller cousins, the damselflies, really pushed their way into my consciousness. James and I had moved to a place of our own, in the village of Groombridge in Kent – and also in East Sussex. The pretty river Grom bisects the village and marks the county border. We lived in the Sussex half. Several mornings a week I would get up at dawn to patrol my 'patch' – a piece of mellow Wealden countryside at once both anonymous and gorgeous – and document the birds and (at the right time of year) butterflies I saw in the fields and woods.

It was late May and I was at the far edge of my patch. I'd had a wonderful morning. It had begun with a look through a hedge at three fox cubs playing together at first light, and things had only got better as I walked alongside a small copse and found myself joined by two badgers – an adult and a smallish cub – trotting home to their sett and completely unworried by my presence. Now I had reached one of my favourite sections. It was a lane through aromatic sheep fields, with lush high hedgerows on either side. I was looking out for Yellowhammers and finches when I saw a flashing flicker of dark shiny insect wings high against the hawthorn leaves. I rushed forward for a closer look but the winged thing disappeared over the top of the hedge.

I stood, bereft and baffled, unable to make sense of what I'd seen. It was an insect, but the only insects I could think of with coloured rather than clear wings were butterflies and moths. I didn't know of many butterflies or moths that were as dark as my mystery creature had seemed to be — and none that were as shiny. I spent a few minutes pointlessly standing where I had last seen the insect, as if I could summon it back with my thoughts, and then reluctantly walked on.

A few strides later, I was again stopped in my tracks by another wing-flicker. This time the wings were not so dark but definitely coloured with a strong golden hue, and I could see that the shape was wrong for any moth or butterfly. The creature settled on my side of the hedge and rather than hurrying towards it I crept up slowly, eyes fixed on the spot where it had landed, until I could make out the full shape of my quarry.

I could see now that it was a damselfly — dainty dumbbell-headed relative of the dragonflies — but it was like no damsel I had ever seen before. For a start, it was big — as long-bodied as a small dragonfly, though much more lightly and delicately built than any dragon. Then there were the colours. From head to abdomen-tip it was an iridescent bronzy amber-green, with long and deep-based golden wings pressed tight to its body. Brooding upon its leaf, head tilted down and backside held aloft, it looked like some metallic mythical beast, or a piece of living jewellery. I had no idea that such wonders existed anywhere, never mind a half-hour walk from where I lived. I wondered if this creature was connected to the dark-winged insect I'd seen before, and very soon I had my answer when the original creature, or another just like it, bobbed gracefully over the top of the hedge, landed close to the bronze damsel and began seductively flicking those glossy violet-black wings.

The dark damsel was, if anything, even more stunning than the bronze one. Its lustrous wings shone with both blue and coppery tones, and its body was the most beautiful deep reflective ultramarine. Watching its behaviour, which involved slowly opening and then quickly flicking shut those fabulous wings, I concluded that it was a male and was displaying to the other, a female. When the two damsels took flight, they flew with tremendous elegance, flitting lightly with deep strokes of their big glossy wings. They went over the hedge together, and despite searching I found no more. Back home I pulled out the trusty *Reader's Digest* book and identified them as Demoiselle Agrions or, as they are known today, Beautiful Demoiselles.

AN EVOLVING OBSESSION

Quite a few years — and one or two more boyfriends — have gone by since the magic self-repairing dragon and the mind-bendingly beautiful damsel made their respective impressions on me. I had slowly built a little more knowledge of the British Odonata, and seen a few of them. Some of the more distinctive species I could identify at a glance — Banded Demoiselle, Brown Hawker, Black-tailed Skimmer (as long as it was an adult male ...). But the many varieties of identical-looking small blue damselflies still left me puzzled. I constantly failed in my attempts to identify the hawker dragonflies that seemed to fly at 100mph and never stop.

Late in 2010 I decided that I would make the following year my year of dragon hunting. I would travel around and see as many species as I could, and in the process learn how to find them, how to identify them (even the females), and become familiar with the way they live their lives. I would also do my best to take half-decent photos of them.

It was a good plan, but I couldn't do it alone, mainly because I didn't have a car. While I do as much wildlife watching at sites accessible by public transport as I can, I knew there would be some places for which a car would be essential. So I recruited the help of Rob.

Rob and I had been together for five years at the start of Project Odonata, and over that time he had showed great indulgence for my obsessive interest in everything wildlife. While he didn't share my passion, over the last couple of years he had developed an equally violent passion of his own — for photography. In particular, he was drawn to the photography of fearsomely challenging subjects. Wildlife fitted the bill nicely — particularly fast-moving and rare wildlife. By 2010 our interests had converged to some extent, to the point where I had a DSLR camera of my own and Rob could tell a Buzzard from a Kestrel, although I remained 'wildlife watcher with a camera' and he would always be 'photographer who likes wildlife'. The idea of chasing dragons for a year appealed to him, and so the plan was a goer.

YEARS OF THE DRAGONS

This book begins with a general look at the dragons and damsels, explaining what they are, how they live and their relationship with humans. The technical details of how to watch them and, in particular, take photos of them are covered in the Appendices. The rest of the book documents in detail the dragonfly and damselfly encounters Rob and I enjoyed through 2011 and 2012. The former was a good year. We saw and photographed the majority of dragonflies and damselflies that occur in Britain. We enjoyed it so much that we decided to do it again in 2012, in the hope of seeing those species that had eluded us in the previous year. Unfortunately, the weather gods conspired against us. After dire warnings in early spring of drought conditions, the late spring and most of the summer of 2012 went very much the other way, with rain, wind and low temperatures for weeks on end. It was an inconvenience for us, though of course much more serious for the wildlife itself.

Despite all this, we had quite a lot of success in 2012, seeing several of the species we'd missed in 2011. Our final tally is missing just three established UK species, two of which we did try to find but without success. Also missing are nearly all of those species that are on the British List but as wanderers from other lands or as 'putative colonists', but we did succeed with one very special uber-rarity. Some of the best parts of the journey, though, were times spent watching any old species, but at leisure, with time to really observe and grow to understand the insect and how it lives.

It wasn't just the weather gods that had it in for us in 2012 — the relationship gods also conspired to throw a spanner in the works and Rob and I went our separate ways in the late summer. However, we remain friends and share many happy memories of our time together with the dragons. I also made a few trips with other friends, fellow wildlife nuts who entered into the spirit of things with great enthusiasm, and a few outings were solo ventures.

I had perhaps expected, after two summers devoted to dragons and damsels, that by the end of it I'd be pretty much sick of the sight of them. That's not happened. As I write, it's nearly September and I'm bringing together the last threads of this book. However, I'd much rather be outside, wandering round my local patch and looking for the late-summer species, the likes of Migrant Hawker and Ruddy Darter, common dragons I've seen hundreds of times now but have never tired of watching. In fact, if it wasn't raining (again), that's probably where I'd be right now. And as for 2013, I may not be writing about it but it's a safe bet that I'll be spending a fair amount of it in much the same way as the last two summers — watching, learning about and enjoying our wonderful dragons and damsels.

CHAPTER 1

The Makings of a Dragon

ᘓC ᘓᘓ

Dragonflies are the attention-seekers of the insect world. Big, bright and brave, they will appear like magic around your newly dug garden pond, sun themselves on your picnic table and fly at you fast enough to make you yelp, duck or both. They live a double life, transforming from skilled swimmers in infancy to matchless fliers in adulthood, in one incredible metamorphosis. They are formidable fighters, deadly hunters and have incredibly complex romantic lives. It's no wonder that almost everyone who loves nature will, sooner or later, fall under their spell. As for damselflies, they may look little and fragile next to their dragon cousins, but they too are fierce and frisky, as well as drop-dead gorgeous.

Dragons and damsels form one of the most easily recognisable of all insect groupings. They are perfect little flying and fly-catching machines, their exquisitely balanced, long-tailed shape recalling a helicopter or biplane, their sturdy legs doubling as landing gear and hunting equipment. Their slim, colourful bodies, the two pairs of powerful wings and the extremely large eyes all add up to make a unique and distinctive look. They have other, less obvious attributes that set them apart from the rest of insect-kind, including bizarre feeding apparatus in their early life stages,

 and the incredibly convoluted way that they mate, which is one of the real puzzles of natural history. In this chapter I'll look at the physical attributes of dragonflies and damselflies – the traits that make them what they are, and how they differ from other insects.

To see where dragons and damsels fit into the grand

scheme of nature, we need to use some basic taxonomy — the rules by which living things are classified. The animal kingdom is divided into several phyla (singular — phylum). Our own phylum is Chordata — creatures with spinal cords. The insects belong to the phylum Arthropoda, which means 'animals with jointed legs'. Alongside the insects in Arthropoda are the crustaceans (crabs, lobsters, barnacles, woodlice and the like), the arachnids (spiders and scorpions), and a few other groups of leggy, scuttling creatures.

Each phylum is further divided into a number of classes. Within Chordata, we humans belong to the class Mammalia — mammals. Within Arthropoda, the insects form a single class called Insecta. That's easy enough to follow so far. The next level down is the order — each class contains several orders. The class Mammalia includes the orders Carnivora (meat-eaters like cats, dogs and bears), Chiroptera (bats) and our own order — Primates. The class Insecta includes such orders as Coleoptera (beetles), Lepidoptera (butterflies and moths), Hymenoptera (bees, ants and wasps), and the one that we are interested in — Odonata (dragonflies and damselflies). Within Odonata the dragonflies form one distinct group (the sub-order Anisoptera) and the damselflies another (Zygoptera).

The Odonata belong to the more primitive subset of insects that do not go through a pupa stage as they develop towards adulthood. Instead, the immature dragonfly or nymph grows to its full size through a series of skin changes, and when it is fully grown it leaves the water and breaks out of its skin one last time as a winged adult. 'Primitive' in this context should not be taken to mean 'not as good as it could be'. The Odonata lifestyle and body form has been a successful formula for many millennia, and the modifying process of evolution tends not to fix that which is not broken. The name 'Odonata' means 'toothed ones', a strangely imprecise name as many other insect groups have similarly toothed jaws.

Insects all have a pretty similar body plan. An adult insect's body is made of repeated segments, which are fused into three distinct 'blocks':

the head, thorax and abdomen. The head bears the insect's sensory and eating equipment: eyes, antennae, mouthparts of some kind. The thorax is the attachment point for the six jointed legs and, if present, one or two pairs of wings. It also houses powerful musculature for making those legs and wings work. The last body part, the abdomen, is usually the largest of the three, and contains most of the bodily organs, the main part of the digestive tract, plus the reproductive bits and pieces.

The head of a dragonfly or damselfly is dominated by its two compound eyes. In dragonflies the eyes are close together, usually touching along the centre, and the two domes of them form the dragon's round head. In damselflies, the eyes are spherical but widely spaced, producing a dumbbell-shaped head. This difference is one of the simplest ways to tell a dragon from a damsel. The eyes are compound, each one made up of lots (about 28,000) of simple but complete eye-like structures called ommatidia, each with its own lens and nerve, all packed together. Because the eyes of dragons and damsels are so large and take up so much space on their heads, they can see in every direction at once. About 80 per cent of the Odonata brain is dedicated to analysing the information received by those monster eyes. By contrast, the antennae, which are organs of smell, are small and simple. These insects are extremely visually oriented, though in their immature, underwater stages they have proportionately smaller eyes and bigger antennae.

Although a compound eye is made up of lots of repeated structures, it doesn't necessarily appear the same all the way across. Most of the Odonata have two or more colours in their eyes. Darker areas of the eye contain more ommatidia and so give better vision — this is why Odonata eyes are often darker on top (the part of the eye that looks forward) than below, as dragons usually approach their prey in a straight line. In those with light-coloured eyes you will probably notice one or more very dark spots or 'pseudopupils' on the eye's surface — these are the points of most acute vision.

Besides the huge compound eyes, there are three very small simple eyes (ocelli) on the tops of their heads. Unlike the compound eyes, these little eyes don't resolve any detail or colour, but react very quickly to changes in light. If you are trying to sneak up on a resting dragonfly and your shadow falls on it, the sudden change in light triggers the ocelli and instantly warns the insect that something big just got in between it and the sun. It will probably fly off immediately rather than wait around to look properly at what that something is.

Dragons and damsels catch their prey with their legs and eat it by taking bites out of it until it has all gone, in contrast to some other predatory insects which employ other, debatably more yucky methods like inserting a sharp tube and sucking out the insides, or vomiting on their lunch to predigest it before slurping it up. The Odonata eating equipment consists of a sturdy and sharply toothed pair of mandibles that open sideways rather than up and down, plus the smaller maxillae that help to manipulate prey. However, the immature insect (nymph) has a rather different arrangement, and uses its labium (a lip-like structure on the lower part of the head) to catch its prey. The labium is highly extensible and can be 'fired' forwards at lightning speed. It is tipped with spikes, which are driven into the prey's body by the speed and force of the strike. The labium is then drawn back, bringing the prey within mandible-reach. This rather startling anatomical feature helped inspire one of the most bizarre and alarming fictional monsters of all time: the titular alien of Ridley Scott's *Alien* films.

Dragonflies have slimmer front wings than hind wings, while damselflies' wings are all quite similar in shape and size. This difference gave rise to the names of the two groups — Anisoptera (dragonflies) comes from the Greek for 'different wings' while Zygoptera (damselflies) means 'paired wings' in Greek. In both groups, the wings are large and well developed, allowing for strong and agile flight. Even the damselflies, which look at first glance to be pretty feeble flutterers, can stay airborne for surprisingly long spells.

The wings are made of a strong but thin, transparent membrane of cuticle (the same tough material that covers the body), supported by a system of veins which spread and fork, dividing the wing membrane up into cells. The pattern of wing venation is distinctly different between dragons and damsels. The veins contain nerves, and are also key in allowing the wings to expand from their initial crumpled state when the brand-new adult insect has just emerged. In most species, near the upper tip of each wing is a coloured, fluid-filled cell called the pterostigma, which lends weight to the outer section of the wings and so helps with stability in flight. When at rest, dragonflies usually hold their wings out at 90 degrees to their bodies, while damsels usually hold their wings folded against their bodies.

Dragons and damsels have six well-developed legs, which they use primarily for clinging onto things (including each other, during mating) and for catching prey. They are not so good for walking, so you'll rarely see a dragon or damsel using its legs to do much locomotion beyond an awkward shuffle (the long, heavy Odonata abdomen is a further handicap to walking). The legs have three main joints — counting outwards from the body they are the femur, tibia and tarsus. The tarsus is tipped with a claw, which helps the insect hook itself securely to its perch.

The thorax, attachment point for the wings and legs, is a blocky structure, packed with muscles for driving those limbs. It is also often the fuzziest part of the insect. We don't really think of dragons and damsels as being fluffy creatures, not like moths or bumblebees, but with a close look you'll see that the thorax is covered in fine hairs (setae). These help trap heat, and as the muscles need a bit of warmth to work, it makes sense that the thorax should be particularly hairy. It also often bears distinctive coloured markings, in particular the antehumeral stripes. The thickness and colour of these paired stripes on the thorax top can be important in determining the species.

The abdomen is the most striking body part in most species, and is often strongly coloured, or patterned, or both. It is also very flexible in the vertical plane. In damselflies and most species of dragonflies it is slim and cylindrical, though in chaser and skimmer dragonflies it is rather flattened and sometimes also cigar-shaped and tapered, in some others it is fatter at the tip than the base, and in males of many species there is a distinct narrow 'waist' near the thorax end. It is made up of 10 segments, the last one of which is tipped with the anal appendages, gripping structures used during copulation. The male uses his to get hold of the female by the back of her head, as an opening gambit in persuading her to mate with him.

Male Odonata could make a convincing argument for being the envy of the animal world, for they have two sets of sexual organs. The primary set are located on the underside of the eighth abdominal segment, near the tail tip, and manufacture sperm. This is pretty standard for insects — both sexes have their bits at the rear end, and to mate they just need to connect their rears together. However, in an Odonata male the organ that actually transfers the sperm to the female is on the underside of segment two, up near his thorax. Given the length of the Odonata abdomen, this is some distance away from where the sperm is made. So prior to mating, the male has to transfer his sperm from segment eight to segment two, and when actually paired with a female he has to somehow get his segment two to come into contact with her segment eight. This is why mating dragons and damsels have to adopt the extraordinary 'wheel' position, with the male gripping the female's head, and the female swinging her abdomen all the way forwards so her tail tip can reach the underneath of the male's abdomen base.

The female's reproductive equipment is bulkier than the male's, especially the elongated ovaries, which take up much of the space inside the abdomen. This is why female Odonata are usually visibly plumper-bodied than males, and don't have the very pinched-in waists seen in the males of some species. Females are also heavier than males, by a matter of a few milligrams.

CHAPTER 2

The Life Cycle of Dragonflies and Damselflies

ᘏ ᘏ

When I was out and about looking for dragons and damsels, my fo-cus was almost exclusively on the insects in their adult form. This was primarily due to the impracticalities of looking underwater for Odo-nata nymphs, but to understand the ways of dragonflies and damselflies it helps to have some insight into their entire life history, including the months or years of immaturity.

All dragonflies and damselflies begin life as an egg, laid by the female in water (or sometimes on dry land but close to water). In some species the egg is just dropped into open water, but in others it is pushed into the tissues of a plant underwater, the female using her ovipositor or egg-laying tube to 'inject' the egg exactly where she wants it. Eggs of the for-mer category are spherical and are laid in clumps usually with a coating of gooey jelly that holds them together and helps them anchor to plants underwater. This is particularly important when the water is flowing and there is a risk of eggs being washed downstream. The darters and emer-ald dragonflies produce these kinds of eggs, while damselflies and hawker dragonflies use the injection method and lay their more elongated eggs singly. In a few cases the female lays her eggs within plant material above or away from the water line, but when they hatch the nymphs will quickly make their way to water.

The egg may develop and hatch quite quickly — within a couple of weeks — but if it is laid towards the end of summer it goes into a kind of suspended animation known as diapause, whereby it doesn't develop at all for months, and the embryo inside only grows and hatches the following spring. Spending winter as an egg has its advantages. An egg is tougher than a small nymph and so may have a better chance of surviving severe winters. On the other hand, an egg is immobile and unable to do anything to escape from danger.

Whether the new dragon or damsel in waiting hatches quickly or after a diapause, the hatching process is the same, and is triggered when the surrounding water temperature reaches about 15°C. When the tiny new nymph emerges, it is wrapped in a protective membrane, but this splits in contact with the water and the nymph is free to begin its independent life. This happens almost immediately if the egg was laid underwater, but if not then the membrane-clad infant wriggles and bounces its way into the water.

Dragonfly and damselfly nymphs look rather different to each other. The most striking difference is at the abdomen end – damsel nymphs are slim rather than squat, and their tails are tipped with three delicate slender fins, called caudal lamellae, which function as gills and extract oxygen from the water. Dragon nymphs have no lamellae (though both dragon and damsel nymphs have gills on their bodies) and their abdomen segments have spines on the outer corners, giving them a 'rough-edged' look compared to the smooth-bodied damsels. Both kinds of nymphs have six sturdy legs, and on their backs are tiny wing-sheaths, within which the full-size adult wings eventually develop.

The life of a nymph has just one purpose. It has no time for anything apart from eating. The faster it eats, the faster it grows, and the larger it becomes the less likely it is to fall prey to another underwater predator, so eating a lot helps sustain its life in more ways than one. By the time it reaches maturity the nymph of a large dragonfly will have become one of the most formidable little killers in the pond, capable of taking tadpoles and even small fish.

Insect bodies aren't made to stretch. There is little 'give' in the outer covering or cuticle of a nymph, so when it gets too tight in there the nymph has to moult its skin. It starts engulfing water, which causes the cuticle to swell and then split along a line of weakness on the thorax. The nymph then struggles out of its old skin, clad in a new soft and pale cuticle that quickly expands to accommodate its bigger body. This is a difficult time – the little creature is both squishier and more visible than usual, so it is at heightened risk of predation for a couple of hours. However, the cuticle soon toughens up and darkens, and normal life (voracious eating through every waking hour) can resume. As an added bonus, if the nymph loses any of its legs or (in the case of damselflies) tail fins, it can grow them back in between moults.

The stages in between moults are called instars or stadia, and the number that a nymph goes through in its lifetime varies from eight to 18, the exact number depending both on species and on the particular conditions an individual experiences during its life. Tropical Odonata can race through their entire larval lives in just a few weeks, but up near the Arctic Circle the process can take five years or more. That means the nymphs must survive several winters. Only if the water is warm enough can the nymph actively hunt for prey. When things cool down too much, the nymph stops moving and enters diapause – all development shuts down. In this state it can even withstand being frozen solid, though younger nymphs are less resistant to this than older, larger ones.

When it is active, the nymph hunts in a different way to the chase-and-grab method of adult dragons and damsels. It finds itself a hiding place among the layer of detritus on the bottom of the lake, pond or river, and waits for moving prey to come along, which it may detect by either smell, via the antennae, or touch via the fine hairs on its cuticle (feeling moving water displaced by the prey), or by sight. The kinds of things it eats vary as it grows – little damselfly nymphs will catch even littler midge and mosquito nymphs, plus water fleas (Daphnia) and aquatic worms. As they get bigger the nymphs move up to fish fry, water snails and tiny immature amphibians. Other Odonata nymphs are fair game too, even those of the same species. All are caught using the remarkable shooting-out lower lip or labium with its impaling spikes.

Once the nymph is fully-grown, and it has detected that the right time of year has arrived (by sensing increases in day length) it is ready to exit the water. Some of the most important physical changes take place a few days before the 'big split' and emergence of the adult insect. The nymph's gills cease to function and its spiracles, the holes in the cuticle through which adult insects breathe, open up. No longer able to take in oxygen from water, it must dwell at the water's surface and breathe air. Its mouth-parts change, too, with the prey-stabbing labium retracting. From now on until the full metamorphosis, it cannot eat anything. Inside the nymph cuticle, the finishing touches are being made to the winged adult, and eventually the creature will climb clear of the water, usually up the stem of a vertical waterside plant.

The insect picks a good spot and grips on with incredible force. Then the nymph casing splits along the back and the adult pushes its head, thorax and legs out, arching outwards as it does so. It takes a breather when the legs are fully out, resting either with its body projecting outwards from the nymph casing or hanging in a bent-over-backwards, upside-down position while its legs firm up. Then it curls itself forwards and gets a grip on the now empty head compartment of the nymphal casing. Bracing itself there it eases its abdomen out.

Once it has extracted itself, the adult dragon or damsel sits quietly, hanging on to its old empty skin (the exuvia), while inside its body various changes are going on. The tiny, scrunched-up wings gradually expand and stiffen as fluids are pumped along their veins. Once the wings are fully pumped up, the fluid used for this is drawn back in and redirected into the abdomen. This now expands too, becoming slimmer and straighter. The insect's body starts to develop its colours, although in most species it will take several more days before full adult coloration has appeared. At this stage of life, young dragons hold their wings flat and pressed together over their backs, damsel-style, but part them in the usual dragon pose when they are nearly ready to take off.

Emergence is an incredibly hazardous time for a dragonfly. If a predator should come along, it would be completely helpless. There is also the risk that something will go wrong in the process — a part of its body could get fatally stuck, an un-noticed obstacle could get in the way of the delicate, expanding wings, it could lose its grip and fall into the water below, where it would surely drown. Perhaps the luckiest of all metamorphosing Odonata are those that do it in front of an interested human observer, who would help keep potential predators away and (perhaps) intervene if the insect got into difficulties, but through my two summers of intensive Odonata-watching I was never lucky enough to witness a metamorphosis for myself.

The fresh dragon or damsel will make its first flight an hour or two after this process is complete, and it will usually head in the opposite direction to the water from whence it came. It may travel many metres or even miles — there is certainly no guarantee it will ever return to the same pool or stretch of river where it was born. It flies in search of a good feeding ground, where it can spend a few days, or even a couple of weeks, living and eating in peace. Over this time it attains full sexual maturity, ready to throw itself into the hurly burly of mating.

Dragonflies and damselflies both catch other flying insects. They use their legs to snare their prey, opening the legs as they get near the prey and then closing them around it. When searching on the wing or giving chase they can fly vertically upwards or down, can hover on the spot, turn on something considerably smaller than a sixpence, and even fly backwards.

Male Odonata will go back to water when they are ready to mate, knowing that this is the best place to meet the ladies. They find a suitable spot and either fly up and down, waiting to intercept a female or chase off an intruding male, or select a lookout perch from where they survey the scene, ready to make their move when the opportunity arises. They spot females of their own species by their colours and flight styles, but take a broad-brush approach and will sometimes approach and attempt to mate with females of the wrong species, or even other males.

Most Odonata don't indulge in any kind of courtship, but the demoiselles do, performing an eyecatching display whereby the colourful wings are quickly flicked open and shut. Often several males will gather at a particularly good stretch of river and sit together wing flicking and sometimes 'bickering' among themselves for prime position. Any female heading riverwards would be hard-pressed to miss this exhibition, which allows her to compare several potential mates and select the 'best'.

In general, females head for suitable areas of water when they have a batch of eggs ready to be fertilised. They can be reasonably confident that they'll encounter at least one male when they get there, if not before — when the waterside is fully occupied, newly arrived males may wander further from the water in search of females. Once a male spots a female, he flies at her, legs first, and tries to seize her in much the same way that he grabs his prey. If he succeeds in getting her in his clutches he will try to lock his claspers onto the back of her head (actually the top of the thorax in damselflies). The female will probably be willing to go along with all this, but if for whatever reason she is not, she can evade his advances by twisting away or just out-flying him in the first place.

Once a male has 'locked on', the two are connected in the 'tandem' position and can go on to mate. For this to happen, the female must bring her tail-tip up under herself to make contact with the male's second abdominal segment, forming the 'wheel' position. This is an awkward manoeuvre to say the least, so it usually happens after they have settled in a suitably safe place. A pair in a mating 'wheel' will stay like this for up to an hour, though if disturbed (by another male trying to steal the female, or perhaps by a careless human) they will remain in tandem and fly off elsewhere to complete the act. During mating, the male will normally take some steps to displace any previous male's sperm from the vicinity of the eggs before depositing his own, either by pushing it aside or by scraping it out.

Once a female's eggs have been fertilised, she will want to lay them swiftly on or by the water. The male, though, will want to see the process through to the end so he stays in attendance, either still physically attached to her in tandem as she lays, or by following her closely while she goes about her business. In the case of the darters, the male stays attached and helps dislodge the female's egg mass at a suitable spot by swooping forwards so her body swings against the water's surface.

A female damselfly may lay a dozen or more batches of eggs in her lifetime, each one containing a couple of hundred eggs. This very high productivity is a necessity, as the vast majority of those eggs will not make it to maturity — there are heavy losses to predators at all stages. It is a long, hard road to adulthood for a dragonfly and adult life lasts at most a few weeks — a much briefer spell than the underwater nymph stage. But when you watch a dragon racing up and down its chosen patch, fearless and fabulous in its aerial mastery, all the danger and hard work it has been through to get to that stage really does look to be worthwhile.

CHAPTER 3

Dragonflies and People

ꩡ ꩡ

Dragonflies are impressive rather than pretty, but a straw poll of my non wildlife-obsessed friends revealed that most do seem to quite like them, though might prefer to keep them somewhat at arms' length. Damselflies, with their more manageable size and disinclination to actually fly at people, won pretty much universal approval. That really isn't bad as human attitudes towards insects go – in general, if it's not a pretty butterfly or a helpful bee, we don't want to know.

Several friends had stories to tell of dragonfly encounters. One claimed to have been bitten by a large dragon, which I at first dismissed as nonsense but later discovered was perfectly feasible and had to apologise. Another spoke of finding a huge dragon trapped in his greenhouse, and hardly daring to go in there and help the frantic creature find its way out. A friend who was setting up a marquee for a show found a hawker dragonfly that had gone to roost on the clothes of a mannequin – at first glance she took it for a giant and spectacular brooch. There were stories of children being paid 50 pence for dragonfly wings by a notorious local eccentric – presumably destined to decorate something or other. Everyone who'd met a dragonfly at close quarters remarked on how enormous it was, and how beautiful.

Back in Victorian times, the done thing with insects was to catch them in a net and then kill them and pin them to a board. Dragons and damsels were disappointing subjects for this practice though, as the colours that are so vivid in life tended to fade away after death. This is because they are temperature-dependent – even in living dragons the blues and greens can fade to grey when the insect becomes chilled. The blue pruinescence seen in some species doesn't fade away, as it is not an integral part of the insect's cuticle but is secreted on top of it — but this means it is very easily rubbed off. Various preservation methods have been tried out to attempt to preserve the insects' colours, with varying degrees of success. Happily, these days most of us would rather take a clear, bright photograph of a living dragonfly than go to the trouble of catching, killing, preparing and mounting a dead one. Even leaving aside all concerns about ethics and conservation, the process in the latter case is much more laborious, and the result much less pleasing.

Many cultures around the world have their own ideas about dragonflies, and by and large the insects are viewed in a positive light. In Japan, the dragonfly represents strength, courage and happiness while the very shape of the country of Japan was likened by an ancient emperor to a pair of dragonflies flying in tandem. Another emperor became fond of dragonflies after one swooped in to devour the mosquito that had just bitten him. Today, Japanese children like to catch and play with living dragonflies, trapping them by the ingenious method of throwing a pebble wrapped with a hair in their flight path for them to catch. The dragonfly mistakes the fast-moving pebble for an insect, attacks it and (maybe) gets tangled up in the flying hair and pulled down to the ground. Elsewhere in East Asia, dragonflies and their nymphs are caught not for fun but for dinner.

The old American name 'devil's darning needle' describes the pointy-tipped shape of some dragonflies' bodies, and legend has it that dragonflies will sew shut the mouths, eyes and ears of naughty children. It's one of those stories devised to keep kids in line but really only stands to create a deep and inappropriate fear of dragonflies. Another misleading name is 'horse-stinger'. Dragonflies will be attracted to horses, and in their presence the horses may twitch and flinch as if stung ... but the dragonflies are attracted to smaller flies swarming around the horses, and it is these flies' bites that upset the horses, not the dragonflies.

Native Americans have many dragonfly-related myths and ideas. The Zuni Indians of New Mexico tell a charming story whereby a boy makes his sister a model dragonfly from corn husks, to cheer her up after they were left behind when their family moved to new lands after a bad harvest. The toy comes to life, and the living dragonfly appeases the 'corn maidens' who bring food for the children. To other American Indians the dragonfly is a powerful symbol of change, because of its metamorphosis from aquatic nymph to flying adult.

In Britain today, dragonflies have a following of keen wildlife-watchers like never before. The British Dragonfly Society has some 1,500 members, and its website hosts much lively discussion about Odonata of all kinds. Many of the keenest dragon-hunters are birdwatchers looking for something new to see. The society's work at furthering our understanding of dragonfly distribution, biology, population and migration habits is invaluable for helping to conserve all Odonata species in Britain so that future generations can enjoy them.

CHAPTER 4

Dragonflies in Britain

ରୁ ରୁ

Here in Britain we don't have that many regularly breeding drag-onflies and damselflies, though if you add on the irregular visitors – the wanderers and vagrants from further south, east and, in one case, west, the number grows quite a bit. Some species are very common and widespread, others extremely scarce and localised. What determines whether a species lives everywhere or hardly anywhere is a combination of how exacting its habitat needs are, and how well it can cope with different temperatures.

In common with all insects, the Odonata as a whole have a southerly bias in Britain. Warmer summers and milder winters just make life a bit easier for the standard dragon or damsel, though there are a few which are adapted to more northerly climes. The species that do best are those that can, as nymphs, happily live and grow in all kinds of habitats, from town park lake to rural river and moorland mire pool to garden pond, and those that, as adults, are good at dispersing from their birthplace and colonising new habitats.

Many of our rarer Odonata are on the northern limits of their total geographic range here – go further south on mainland Europe and they are much more numerous. However, the general trend to a warmer climate is making Britain ever more hospitable to southerly species, and we have seen several new arrivals begin to colonise parts of southern England in recent years. The flipside of this, of course, is that our northerly specialists – the likes of Azure Hawker and Northern Damselfly – may begin to find Britain just too warm for them.

In this chapter I'll be giving a brief overview of Britain's dragonfly and damselfly fauna, going through the main families, their particular traits, and the habitats they use. The order more or less follows the standard taxonomic order, working from the more primitive damselflies up through the dragonflies to end with the most advanced (evolutionarily speaking) — the darter dragonflies. By 'primitive', I mean those species that retain more ancestral features, while the more 'advanced' species have a higher proportion of newly evolved features. But it is certainly true to say that all species are equally 'evolved', as they have all successfully made it to the present day, and so dealt with and adapted to whatever trials of life have been thrown at them.

It's a bit ironic that the most primitive of our Odonata is a group often considered the most spectacular and beautiful. These are the demoiselles, which rejoice in the family name Calopterygidae. They are the largest of the damselflies, with highly coloured metallic bodies and partly or wholly coloured wings as well. We have two species, the Banded and the Beautiful, and both are absolutely stunning.

Demoiselles need running water to live and breed. The Banded likes slow, lazy and lush rivers, while the Beautiful prefers faster-flowing water. They can be seen gathered at the riverside on warm days from mid-May, the males displaying at each other with wing-flicks, and courting any arriving females with a fluttering flight 'dance'. Females of both species lay their eggs directly into underwater plants. When not actively seeking to mate, they may be found some distance from water.

The Banded Demoiselle is the commoner of the two, living in most of England and Wales, and also Ireland, and just edging into Scotland. The best places to look for it are around lowland rivers with lots of bank-side vegetation. The Beautiful Demoiselle has a more patchy distribution, with a south-westerly bias. It is absent from most of East Anglia, northern England, Ireland and Scotland, but does have some isolated colonies as far north as the Isle of Skye.

The emerald damselflies, belonging to the family Lestidae, form a distinctive group. They are also known as 'spreadwings', because unlike most damselflies they habitually rest with their wings held away from their bodies. In Britain there are two regularly breeding species, two more that seem to be in the early stages of colonisation, and one that has been recorded only once, as a vagrant.

Emeralds get their name from their bright metallic green bodies, sometimes accessorised with a touch of powder blue in the males. They are relatively large damsels, though not as big as the demoiselles. All of the species show a preference for shallow waters with plenty of marginal vegetation, but our commonest species, the Emerald Damselfly, seems to be the most versatile, and also happy to cope with quite an extreme temperature range – it is found almost everywhere in Britain where there is suitable habitat, right up to the far north of Scotland. If you are very lucky, it might even live around your garden pond.

Of the family Platycnemididae, known as 'white-legged damselflies' or, more poetically 'featherlegs', we have only one species – the White-legged Damselfly. This is a medium-sized damsel with several distinctive traits: in particular the wide and frilly white legs and a particularly broad head. It is not a very common species, being restricted to southern England and west Wales where it is patchily distributed. It has fairly specific habitat needs, liking slow, undisturbed, sunny riversides and canals with plenty of emergent vegetation. However, in an optimum habitat, it can be amazingly numerous.

The next damselfly family is the Coenagrionidae. It is very large, with many representatives in Britain, and includes our most familiar damsel species, the ones you're most likely to bump into on an average country walk somewhere with still or flowing water, as well as a few real rarities. They are mostly pretty small, blue damsels, but the group also includes our two bright red species.

There are three damselfly species that you will see with ease almost everywhere in Britain, though one of them does become scarce or absent in the far north of Scotland. If you have a pond, they may well colonise it, but equally you are very likely to find them while you're wandering far off the beaten track. On my dragonfly and damselfly searches up and down Great Britain, there were two species that were literally everywhere that I went, from the parks and gardens of the Home Counties to peatland pools in the bleak moorlands of the western Highlands. These were the Large Red Damselfly and the Common Blue Damselfly. The other very common and widespread species is the Blue-tailed.

All of these damsels emerge early. I saw my first individuals of all three species in week 1 of the dragon-hunt in 2011, in April (though this was very early for Blue-tailed) and was still seeing them all well into August. None of the three species was ever overwhelmingly numerous at any one site, but their apparent ability to successfully live and breed anywhere at all means that they happily live alongside large populations of other, more specialised damselflies in all sorts of habitats. The Large Red is arrestingly red, the Common Blue is strikingly bright blue, and it's hard to miss the Blue-tailed's bright blue tail-tip, on its otherwise mostly black, needle-slim body. Three very common and very different damsels: if there were no other species in Britain, identification would be an absolute breeze.

There are many more, though. The only species that can be confused with the Large Red is the much scarcer Small Red, but among the blues and the blue-and-blacks there are nine more to worry about. Most have very local distributions, but the Azure Damselfly is very common and widespread everywhere apart from Scotland. Learning to tell Azure from Common Blue is one of the first hurdles a beginner Odonataphile needs to cross. This actually isn't that difficult once you know exactly what to look for, but the other blues all present tougher identification challenges.

The rarer Coenagrionidae species generally have quite narrow ranges of habitats they will use, or are limited by climate factors. The Small Red Damselfly, for example, only lives around boggy, acidic pools, while the

Scarce Blue-tailed Damselfly likes small, shallow pools with little vegetation. In terms of climactic limits, most of the localised species are missing north of the Humber, with Northern and Irish Damselflies the notable exceptions. The former species lives only in the Scottish Highlands, while the latter is found only in Ireland and not Great Britain at all (despite occurring elsewhere on mainland Europe). Two of the Coenagrionidae became extinct in Britain in the 20th century, though one of them may now be re-colonising.

The two red-eyed damselfly species stand out from the rest of the blue Coenagrionidae species by virtue of their rather striking blood-red eyes (greenish in females). These damsels are distinctly different in both appearance and behaviour to the rest of the group. Their penchant for sitting around on lily pads makes them easy to find in the right habitat, particularly lush lakes and slow-flowing rivers.

The first of the dragonfly families is Aeshnidae, the hawker dragonflies. These are mostly large (some very large) with proportionately long, straight-sided abdomens. They usually have vivid blue or green markings on a dark body. Eight species breed in Britain regularly, with four more making infrequent appearances as wanderers from abroad.

Aeshnidae species are the archetypal dragonflies. Big, colourful, long and sleek, they are the ones that zoom past your ear with millimetres to spare, their wings making a startling rattling sound as they go by. You can wait all day for one to settle for a photograph – they seem to have boundless energy. They fly with consummate skill and panache, casually plucking smaller and slower insects out of the air and consuming them in flight, discarding the prey's pathetically tiny wings like sweet wrappers. When they do land, usually high in a tree, they dangle rather than perch. Most of the eight species that breed here have a southerly distribution, but the Common Hawker is more common in the north, the Norfolk Hawker is restricted to East Anglia, and the Azure Hawker lives only in Scotland.

The UK's resident Aeshnidae are associated with still, usually large and low-lying waters, but you may encounter individuals a long way from water, as they tend to move to woodland or open countryside areas for the period (a week or more) in between emerging from the water and attaining sexual maturity. The vagrant Aeshnidae on the British list include the only dragonfly to have made it here from North America – the Common Green Darner. One of the other three, the Lesser Emperor, may be poised to colonise Britain, as sightings are on the increase.

The family Gomphidae or club-tailed dragonflies has only one resident British representative, the rather rare Club-tailed Dragonfly or Common Clubtail. It, and its family as a whole, are named for the body shape which swells to a broad, club-like tip. Unlike most dragonflies, its eyes don't meet in the middle but are clearly separated, and proportionately smaller than the eyes of other dragonflies. One of the few British Odonata which needs flowing water to breed, it lives around large, slow rivers, mainly in southern England and Wales. Another clubtail species, the Yellow-legged Dragonfly, occurred once in Britain, nearly 200 years ago.

Another family with just one British representative is Cordulegastridae. This family, the golden-ringed dragonflies, are exactly as their name suggests – marked with prominent gold rings on otherwise dark bodies. The Golden-ringed Dragonfly is a large and stunning species of heath and moorland, with a strongly westerly and northerly bias to its distribution (but it does not occur in Ireland).

The emerald dragonflies, family Corduliidae, have three British representatives – a fourth became extinct in the 20th century. Emerald dragons, not to be confused with emerald damselflies, are medium-sized with metallic green bodies, club-shaped in males but straight-sided in females. They are very active dragonflies, with a fast, erratic flight, and a marked propensity for territoriality. Emeralds generally like good-sized, shallow and sheltered lakes, and all three species are uncommon and localised. Of the three, the Downy Emerald is the most widespread, but its distribution is very patchy.

The last dragonfly family is a very large one. The Libellulidae contains nine resident British species, but another eight have appeared as visitors, and one of those, the Red-winged Darter, keeps threatening to establish itself properly here. British libellulids can be roughly divided into two distinct groups — the chasers/skimmers, and the darters.

In all but one of our chasers and skimmers, the males are blue and the females brown or yellowish. (The exception is the Four-spotted Chaser, in which both sexes are greenish-brown). The males' blueness comes from pruinescence, a waxy secretion that develops gradually with age. They are solidly built with tapered and somewhat flattened bodies, and some have obvious dark patches on their wings. Chasers and skimmers are the photographer's friends, because they spend much time sitting on a favourite lookout perch, often a nice prominent bare stick, and though they do set off to catch prey or chase off a rival from time to time, they return to the same perch again and again. The Four-spotted Chaser is quite possibly the most common and widespread dragonfly seen in Britain, while of the other species the Black-tailed Skimmer and Broad-bodied Chaser are both common in most of England and Wales, while the remaining two — the Scarce Chaser and Keeled Skimmer — are much scarcer. These dragonflies generally prefer still water, and may use quite small ponds.

The darters include our smallest dragonflies, although their round heads and strong build distinguish them from damselflies. In most species the males are red and the females yellowish-brown, although in two species the males are predominantly black. Darters have a fast and unpredictable dashing flight, but they settle frequently, and often return to the same favourite spot, though they are more likely to choose for their perch a large flat surface than an isolated stick. They frequently rest on the ground, on rocks or on bare wood — benches, picnic tables, boardwalks and handrails are all popular.

Of the four darters that breed in Britain, two (the Common and Ruddy) are quite relaxed about habitat choices and live around all kinds of wetlands, the Common Darter occurring throughout Britain, while the Ruddy is more southerly. The other two (Black and White-faced) are fussier and stick to acid heath or moorland, favouring more northern and western parts of Britain. The seven species that don't breed here regularly range from the Red-veined Darter which keeps establishing small but short-lived colonies and may become a 'proper' breeding species here in due course, to the Banded Darter which has only been seen here once. In the year 1995 there was a major arrival of assorted darter species from the continent, including that one solitary Banded Darter and records of several other major rarities.

CHAPTER 5

Anax ephippiger The Vagrant Emperor

ᕼᏜ Ꮪᕽ

I had expected our dragon-watching adventure to begin with something unremarkable — a common species that we'd be seeing again and again, as the spring and summer weeks rolled by. Probably it would be an early Large Red Damselfly, fresh from the water and flittering unsteadily across my sightline as I looked for something else.

Our first intended target was the earliest-emerging dragon, the little Hairy Dragonfly, and I had planned a visit to Oare Marshes to look for it on the first May bank holiday weekend. I liked the idea of heading out on the day of the royal wedding, taking advantage of empty roads and quiet trails. Years from now, people would ask, 'Where were you when William and Kate got hitched?' We would reply, 'In a ditch in Kent, looking at an insect,' and watch them make their excuses and slowly back away.

That's not quite how it worked out, though. Through a quirk of the calendar, Easter fell the weekend before May Day weekend, and we'd already spent most of Easter looking at birds and butterflies — Kittiwakes and Peregrines at Seaford Head, Grizzled and Dingy Skippers on the North Downs. On the Sunday night I was idly surfing the net for ideas of where to go on our last day off when I stumbled upon something wholly unexpected.

The website was Birdforum.net, online meeting place for wildlife enthusiasts of every stripe. The thread that had caught my eye was titled: 'Vagrant Emperors in Kent'. 'Hey, we live in Kent,' I thought. 'I wonder if that's a kind of dragonfly?' I read the thread. Three or four Vagrant Emperors had been spotted at the RSPB reserve at Dungeness. It was a kind of dragonfly. Someone with the username 'MarkHows' had posted two jaw-dropping photographs of one of the males in flight – a powerful-looking brown and amber dragon with a pearly blue 'saddle' on its abdomen, just behind its yellow-tinted wings. It was not a British species and therefore not on our 'to see' list, but that didn't seem to matter – there was no way we weren't going to Dungeness tomorrow.

Here in the UK, we have one resident dragonfly that rejoices in the name 'Emperor'. The Emperor Dragonfly, *Anax imperator*, is a very large and very colourful dragon that usually first appears in June. No other *Anax* makes its home here, but we receive periodic visitations from two others – the Lesser Emperor, and the closely related Vagrant Emperor. *Anax* is Greek for 'lord', 'master' or 'king' (so *Hemianax* means 'half a lord/master/king') and these superb dragons deserve such a noble moniker, though neither the Lesser nor the Vagrant is as quite as splendid as *imperator* (that means 'commander').

That night I read up some more about our quarry. The Vagrant Emperor lives across nearly all of Africa south of the Sahara, and at similar latitudes further east. It is a fast breeder, making the most of a very different climate and seasonal pattern, and takes full advantage of temporary pools that appear after the rains to produce a new generation of adults in as little as three months. This is a far cry from our own Emperor Dragonfly, which usually takes two years to complete its lifecycle.

Every so often, local Vagrant Emperor populations grow too large for the available habitat and a mass migration is triggered. Thousands of the strong-flying insects storm northwards in search of new places to breed. Some cross the bleak expanse of the Sahara, some push on into Europe. A handful may make it as far as the UK – an occurrence that has become gradually more frequent since the first record, in 1903. This amazing flyer has even reached Iceland, a feat unmatched by any other species. A south-westerly wind will help things along – when Saharan dust is carried to our shores, as happens on occasion, and dumped as a fine red mist on our cars, that's a good time to dust off the windscreen and go out to look for Vagrant Emperors.

We were on our way by mid-morning. As we drove, I watched a flock of House Martins spinning around one particular house in a row of five in Borough Green village as we headed for the M20. It was the fifth unseasonably hot and sunny day in a row, and I wondered whether this was why this intrepid trio of African dragons had been compelled to fly so far north, across the English Channel. The weather was forecast to change that week, with winds swinging round to blow chilly northern air our way. What would the pioneer emperors do then? Die from cold? Probably not, given that they have been seen in the UK in the depths of winter. They might just turn tail and head back south, but with their instincts firmly in migration mode, they might carry on northwards. Maybe this time next week there would be three hundred or three thousand Vagrant Emperors in the UK, rather than just three.

Dungeness is a vast triangular peninsula of shingle that juts out into the English Channel and offers landfall to exhausted migrants. It has long been famed for rare birds, and it's easy to see why an exhausted warbler, shrike or pipit, blown off its usual migratory route, would make a bee-line for the shingly shore and find a clump of vegetation to hide in while it got its breath back. But how would this coastline look to a dragonfly? Possibly not very inviting, but just a short way inland the shingle becomes more and more vegetated, with scrub, farmland and plenty of water in the form of pools, creeks and ditches.

Driving along Dengemarsh Road at Dungeness, we could see a crowd of people on a flat stone bridge out at the area of the reserve called Dengemarsh. Their cars were strung out in a long, shiny line on the roadside. We parked at the end of the queue and joined a small group walking bridge-wards. When we got there, one of the men already there gave us a sidelong glance and remarked, 'I wonder if this bridge has a weight restriction?'

Choosing not to take this personally, I found a place to stand, and surveyed the scene.

The bridge crossed a biggish creek, or maybe a smallish river that had steeply sloping reedy banks and glitteringly clear, slow-flowing water. Between the people and the road were fields of crops, and ahead lay the extensive open waters of Dengemarsh, dotted here and there with wildfowl. Sedge Warblers and Whitethroats sang loudly from the small, scattered bushes, and beyond the water a pair of Marsh Harriers lazily dipped and wheeled over the reeds. It appeared an idyllic scene, though not for the people on the bridge, who'd come to see a dragonfly (or three) that was apparently no longer here.

If you're prepared to travel specially to see a specific individual rare bird or birds, in order to increase the number of species you have seen, what you are doing is twitching. You're out on a twitch, and because you're nervous about whether or not you will actually see your target bird, you are probably literally twitching too. People who do this kind of bird watching all the time are known as twitchers. If you call an ordinary birder a twitcher, he or she will probably reprimand you, or at least give you a tight-lipped smile. Twitching isn't the kind of thing most birders like to admit to doing on anything other than a casual basis. Nevertheless, the atmosphere up on that bridge was decidedly twitchy, and even though they were here for an insect on this occasion, the people present were mainly birders. The murmured conversation around us was of Red-footed Falcons and Black Kites. I eavesdropped on one pair talking about how, from an original interest in birds, they had gone on to become twitchers of butterflies, dragonflies and even orchids ... but only in moments where there were no rare birds to chase.

The passion of the twitcher is that of the collector, and the object is to build a collection that outshines all others. Rewind a few decades and we might have been talking about *literally* collecting the bird, butterfly or other creature, whether with shotgun or net, though dragonflies' dazzling colours fade so quickly after death that they must have been disappointing targets for Victorian naturalists. Now, the collector's tool is a camera, and most of those present – including Rob and I – were wielding some kind of Canon- or Nikon-branded glassware, along with the more traditional binoculars and telescopes. A photograph of the living creature was the prize now, but for us at least a mental snapshot would do – that moment when you see your quarry clearly and unmistakeably through the view-finder of your eyes.

Down in the creek, I saw a movement near the bank and focused on a smallish, bright green dragonfly picking its way along the fringing vegetation. The man next to me saw it too and said, 'Hairy,' in disappointment. It was a female Hairy Dragonfly, a compact brown-eyed dragonfly with gleaming green pearl-spots along a short and slim dark body. Unlike the large hawker dragons, which race imperiously along at head-height and breakneck speed, the Hairy Dragonfly was dipping into and out of the vegetation, making slow and steady progress upriver and out of view.

I'd last seen a Hairy Dragonfly more than 15 years ago. It is an easy dragonfly to identify, because it is on the wing a good month before any of the true hawker dragonflies, and is a lot smaller than them. It shares with the other hawkers a restless disposition, tirelessly patrolling up and down its patch as long as the air temperature is warm enough. You could wait for hours to see one settle and still be disappointed. The view through binoculars reveals that its long straight body is dark but marked with rows of apple green or sky-blue spots – more green on the females, more blue on the males.

'What's that big fish?' asked the same man who'd identified the Hairy. I looked where he was looking. The fish had a curiously front-heavy look, with glittering, sinister bright eyes set high and well back on the sides of its large, shovel-shaped head. 'Pike,' said someone else. A ripple of interest ran through the crowd. Several heads craned over the bridge, and there was a sharp 'ping', muffled splash and not so muffled swear word as something plastic and expensive fell from a camera, bounced off the steel grille below us and into the water. Unperturbed, the Pike regarded the humans' antics with cold disinterest, and slowly drifted out of view.

Very quietly, I explained to Rob how to tell a Hairy Dragonfly from a Vagrant Emperor, as I didn't want to reveal our ignorance to the very clued-up crowd around us. 'Right. Got you. Not green,' said Rob, *sotto voce*. Then, as I went for a wander along the riverbank, a couple pointed out a lovely surprise — a Hairy Dragonfly resting in the reeds. This was my chance for a close look at the species, and an opportunity to take a few photos.

The dragon was clinging to a vertical reed leaf, its neat long legs all bunched together and its claws hooked around the leaf in a tight embrace. Its wings were clasped shut above its body, in a posture that is rarely seen in dragons except when they are very sleepy or freshly emerged from their nymphal casings. This Hairy's posture and subdued colours suggested to me that it was experiencing its first morning of adult life. Its immense, dull blue eyes were blank as a switched-off TV, and bore a soft violet-pink lustre. At this range, I could even see the cloak of fine fuzz on the thorax that gives the species its name. Warming under the sun, it bore our close examination and clicking cameras without any sign of annoyance. After a few photos, I headed back towards the bridge. I glanced over my shoulder halfway there and saw the young Hairy take off from its riverside perch, flying with panache and confidence down the creek in search of its first meal.

Although the Hairy Dragonfly is sometimes called the Hairy Hawker, it is in a different genus to our 'true' hawkers, and in fact is the only species in the world to be placed in the genus *Brachytron*. The meaning of this Greek-derived name, rather embarrassingly for the male Hairy Dragonflies, is 'short tool', though this appears to refer to the dragon's overall body length (it is smaller than any of the 'true' hawkers') and not to the length of its claspers, which are in fact proportionately quite long. Its species name, *pratense*, is from Latin and means 'of the meadows' — an odd name for a dragonfly that is usually seen hawking over water. Its British distribution is patchy but mainly in the south-east (plus most of Ireland). It is on the increase, its main population core pushing north and westwards towards the few isolated outposts in northern England and Scotland.

By now, the crowd was spread out along the river, with a small hard-core group waiting on the bridge. People were beginning to give up and wander off, amid theories that the Vagrant Emperors had flown on north, or been driven away by territorial Hairy Dragonflies. I took some photos of a nearby Sedge Warbler as he belted out his chaotic mess of a song while twirling gracefully around an upright twig like a skilled pole-dancer. A Blue-tailed Damselfly, impossibly delicate-looking, lifted up on a breeze and down again to the water's edge — our second species of the year, but not the one we wanted right now. Rob had sat down on the riverbank and I decided to join him to discuss what we should do next.

I was halfway towards him when I saw that the people further down the river were running away from where we were. I looked back to see that the few people on the bridge were running too, pushing through the gate and down onto the riverside. Suddenly we were caught in an exodus of anxious twitchers, trying to cover the rough ground as fast as they could, while controlling wildly swinging optics.

We screeched to a halt at the riverside just in time to see a golden-winged dragonfly nipping around the bend away from us. We ran on, it flew on. For a few seconds we could view it before it flitted out of sight again around the next bend. It was enough to see the blue 'saddle' at the abdomen base that told us beyond doubt that we were looking at a male Vagrant Emperor. This distinctive trait explains the species' scientific name — *ephippiger* means a horse's saddle.

Finally, the dragon turned and headed back downriver. We all watched as it flicked past, fast and purposeful, showing off its unique pattern and the gold sheen of its wings. Shutters clicked, then it was gone and we were chasing it again, but with less panicked urgency than before. Rob caught up with me on the return run, and then somehow it seemed silly to keep chasing this dragon up and down, so we sat down on the bank and waited. Sure enough, it was soon zooming past us again, and we both fired shots at it. I knew I'd failed to get a photo of anything other than a brownish blur. 'Get any shots?' I asked Rob. 'Well,' he replied. 'I got something. I think.' He tried to show me, but the sunshine was too bright for the image on the camera's screen to show up. We would have to wait.

The dragon did another flyby, and this time I just followed it with my binoculars. Like a tiny helicopter, it flew with its front end tilted downward, fast and furious with lightning direction changes. There was no shortage of small flying insects around for it to hunt, from tiny midges to more substantial fare like the fuzzy black St Mark's Flies with their dangling forelimbs and bobbing flight, or the plump and bumbling alderflies that were festooning the reeds and dragonfly-watchers alike.

All over the UK, pagers and phones would be beeping in pockets, announcing the news of the Vagrant Emperor. Some of the people who had left earlier were already returning, hurrying along the bank to get their look at the star of the day. Soon more would arrive. However, after the fourth flypast we waited and waited but it seemed that the show was over. There were rumours that the dragon had been seen heading away from the creek and across the fields.

Would it return? The whole Dungeness peninsula is crisscrossed with creeks just like this one, many of them on private land. A Vagrant Emperor (or three) could easily get lost in this wide and watery landscape. Bad news for the would-be dragon-watchers, but good news for the insect itself — it was hard to imagine a better place for a dragonfly to rest and eke out its short life, especially after such an epic flight from its native lands.

We stayed put a while longer, just enjoying the place, though I was still eager to know whether Rob had managed any good photos. Down in the creek, a damselfly landed daintily on a short stick poking out of the water. It was a male Common Blue. Its powder-blue abdomen stuck out at right angles to its perch and was neatly and finely banded with black, with the characteristic black 'rugby ball on a stick' marking on the second abdominal segment. It rested there a while, meditatively flexing and straightening its delicate needle of a body, perhaps hoping for a female to happen along.

Heading back along the road, Rob noticed a disastrously flat rear tyre on one of the parked cars. He waylaid the driver when she returned, and after a brief conversation, put on the spare wheel for her with impressive efficiency. Surplus to requirements, I went back to our car and sneaked a look at his photos in the shade of the back seat. There was blur after blur, but among the blurs was a short sequence of sharp images of the Vagrant Emperor flying towards the camera, almost head-on, its great greeny-brown eyes staring me down, the diagnostic blue saddle on its narrow back showing clearly between the four filmy golden planes of its wings.

CHAPTER 6

Libellula depressa
The Broad-bodied Chaser

❧ ❧

Two days had passed since we struck lucky with the Vagrant Emperor at Dungeness. Delighted though I was with such an unexpected bonus right at the start of the quest, we were already lagging behind the country at large. As I explored various websites I found that people were already seeing Broad-bodied and Four-spotted Chasers, Large Red and Red-eyed Damselflies, and the first Banded and Beautiful Demoiselles. None of these are rare or difficult species, and they would all be around for weeks or months to come, but it was hard to escape a nagging sense of everything passing us by.

Despite this, I didn't have dragonflies in mind when I went to see my friend Susan on the Wednesday in between Easter and May Day. As she has a big garden, I always take my camera when I visit her in the daytime, and I spent a happy couple of hours taking photos of tulips and butterflies while she got on with some gardening. Temperatures had fallen to a more reasonable 18 or so degrees C since the weekend, but it remained beautifully blue-skied. At about 4pm Susan suggested a walk and we wandered down the lane to a footpath that traversed a straggly old apple orchard.

We were in Pembury, a villagey offshoot of Tunbridge Wells, in the part of Kent known as the High Weald. A far cry from the wide beaches and ironed-flat pasture of the Dungeness area, this area is hilly and generously wooded. From the path through the apple trees we looked down into a dell brimming over with mature deciduous trees, all coming into leaf. Around us, Whitethroats sang their brief, sore-throated ditties, some hiding out of sight and others performing an exuberant up-and-down songflight.

In a sheltered spot, Susan threw down her chequered blanket and lay down for a doze. I carried on downhill towards the tall trees, eyes and ears on full alert for birds, butterflies, everything. Once I was close to the trees, I picked up a sound so subtle that I had to stand still for minutes, straining my ears, trying to filter out the sounds of thousands of leaves brushing against each other in the wind to home in on a very soft and tentative tapping from a dead branch above me.

Persistent tapping against wood usually means a woodpecker or a Nuthatch. I found the bird in question but it was almost hidden on the far and shady side of its branch. Through the camera's viewfinder, I could see the silhouetted outline of its head as it drew back for the next peck, chipping away at the dead wood in search of some juicy insect larva within. It was a small and dainty head, round and short-billed. I knew what I thought it was, but it wasn't until the bird finally finished what it was doing and flew to a better-lit branch that I saw the black-and-white ladder-striped back that confirmed it was a Lesser Spotted Woodpecker — smallest, shyest and scarcest of our woodpeckers.

The few photos I managed before the little bird had briskly shinned up out of sight among the dense leaves were terrible. It didn't matter. The encounter felt like an immense privilege — a few seconds in the company of some ephemeral woodland fairy or phantom. Then I heard its high, agitated call from deeper in the trees, and turned back to brag about my sighting to a sleepy Susan.

We walked back down the hill, and I pointed out the spot where I had seen the woodpecker. It was long gone. We went down a steep grassy slope to follow the path into the trees, but I was stopped in my tracks by a Comma butterfly, gliding on spread fire-coloured wings around a tall hawthorn. It alighted delicately on a high leaf and sat there pertly, antennae erect and quivering as it monitored its territory, ready to dive off and chase away any intruders. On the other side of the hawthorn were other Lepidoptera. A veritable swarm of tiny blackish moths were spinning around the leaves, flaunting extraordinarily long white antennae that danced around them like delicate ribbons. When one landed, its neat little wings reflected a golden sheen, and those outrageous antennae — at least six times the length of the moth itself — fluttered in the breeze. These moths were *Adela reaumurella* — I only knew this because I'd met them before, a week ago, while chasing butterflies on the North Downs. It seemed extraordinary that such a striking little insect had no English name, though the family Adelidae does have an appropriate moniker — the fairy longhorn moths.

While I was looking up, Susan was looking down, and gave an exclamation of mixed excitement and alarm when she spotted a Hornet investigating the flowers around the base of the hawthorn. Another soon joined it. The two huge, vividly tawny-yellow wasps evaded my camera, but Susan was on top wildlife-spotting form and asked me to identify another big yellow insect that she'd just seen settling on a dead bramble stalk. I saw at once that it was a dragonfly.

This Broad-bodied Chaser, for that was what it was, actually looked like a huge wasp at first glance, with its cigar-shaped yellow abdomen. Some experts think that the resemblance is no coincidence, and that the Broad-bodied Chaser females and immature males have evolved to look like Hornets. The advantages of looking like a Hornet are obvious — many predators have already learned not to approach an insect that carries a large and painful sting. The phenomenon of harmless animals looking like dangerous or bad-tasting ones to benefit from predators' learned avoidance of the latter is common in nature, and is known as Batesian mimicry.

The trick only works well if the 'model' (the dangerous animal that's being imitated) is quite common in the environment that the two species share — common enough that the local predators meet it at least as often as they meet its harmless (and delicious) mimic. That may explain why we don't see very many examples of Batesian mimicry in dragonflies and damselflies — there aren't many potential models around that are common enough, large enough and close enough to the basic Odonata body shape to work for them. However, things may be different overseas — for example, there's some evidence from the tropics of South America that certain damselflies are Batesian mimics of horrible-tasting butterflies.

Mature male Broad-bodied Chasers develop a distinctive powder-blue pruinescence with age, but females and fresh males alike are golden-coloured, darkening towards the tapered abdomen tip and marked along the sides with yellow ovals. On examination of my photos later, I decided that this one was an immature male, because of its rather pointy rear end and longish anal appendages — the paired 'claspers' that it would use to firmly grip its partner's neck prior to copulation. For now, the young dragon was content to rest on its perch, giving us the chance to admire the tracery of veins across its immaculate wings.

The second part of the scientific name *Libellula depressa* describes the curiously squashed or pressed-down appearance of the Broad-bodied Chaser's body. ('Libellula' itself just means 'dragonfly'.) From above it looks wide-bodied, almost portly, but from the side it looks flattened. This distinctive shape readily distinguishes it from the other *Libellula* dragons, but it shares with them the habit of adopting a favourite perch and returning to it again and again. From this vantage point, it watches for the movement of potential prey, and dashes out to intercept anything that looks edible. In this respect the chasers are quite different to the restless hawkers, that favour a ceaseless aerial patrol when hunting. This also means that chasers are much easier to photograph than hawkers. I took photo after frame-filling photo of our resting chaser, tweaking settings to adjust my depth of field, to try to get every millimetre of it in sharp focus. When we carried on, the path took us across a tiny stream — not really a perfect breeding ground for the species, but it could have flown here from a more distant slow river or lake.

Pick any footpath in the UK in spring or summer and you could bump into a dragonfly or damselfly. Though all species depend on water to complete their life cycles, adults may wander some distance away from their birthplaces, with the larger and more powerfully flying dragons the most likely to be found in surprising places. I once noticed a large, bright green dragon zooming the wrong way down a busy one-way street in the centre of Tunbridge Wells. At the time, I decided it was a Brilliant Emerald, though now I'm not so sure — and I don't think I'll be heading back to that street to look for another one. Wandering is a risky strategy — the dragon may not find a mate or a suitable breeding place — but on the other hand it could discover a rich and underexploited new patch of habitat.

The walk back took us along a lovely quiet lane, and we paused to admire a field packed of spring lambs, frolicking in the traditional manner and looking very pretty and charming. Then I realised that we weren't the only ones watching the lambs. At the far edge of the field sat a vixen, positively aglow in the late afternoon sunshine, her beautiful face turned our way. In fact, she was probably watching us rather than the lambs, and as we watched her right back, her nerve broke and she crept away.

The Broad-bodied Chaser wasn't the only new Odonata species for the list that I encountered that week without really trying. On Saturday I went for an afternoon walk with another friend, Michèle, and although she decided the route with views rather than Odonata in mind it did include some watery stretches. By now the northerly wind had picked up and there was a real chill in the air, despite yet more sunshine.

We headed out from Haysden Country Park near Tonbridge. The River Medway flows through the park and sends off numerous tributaries of various sizes, around which the footpaths weave. Walking alongside a little ditch, Michèle pointed out a damselfly sitting unobtrusively on a nettle leaf.

I was starting to feel a little paranoid: almost every dragon or damsel I'd seen so far this year had been found by someone else, who was kind enough to point it out to me. I would have walked right past this one, but thanks to Michèle I stopped for a good look and some photos.

The damsel was a male Large Red, the species that I'd expected to see before any others this year. As it was, it came in sixth — but perhaps I'd overlooked dozens of them in the last week. A biggish damsel, it was a pleasing shade of deep, rich crimson red from head to tail, with bronze panels on the thorax and a couple of blackish segments towards the end of the abdomen. It was in full rest mode, the narrow silvery wings pressed close to its body. Expending energy on a rather chilly day like this was not a good idea. There would be plenty of time to wait for warmer weather and the chance to grab a female Large Red.

The Large Red Damselfly is a very common and widespread species, which is quick to colonise all kinds of wetland habitats. If you see a red damselfly anywhere, there is a strong probability it is this species (as the Small Red is much scarcer). If you have just dug a new garden pond, the chances are that this will be the first damselfly to appear beside it. It also has a long flight season, appearing in April and hanging on into August. It is our only damselfly of the genus *Pyrrhosoma*, an attractive name that means 'flame-coloured body' in Greek. The species name, *nymphula*, is just as poetic, meaning 'nymph-like'. We left the flame-bodied nymph-like little insect in peace and carried on, climbing through wheat fields to Bidborough Ridge, overlooking a luscious landscape of rolling wooded hills.

In sheltered spots, Orange-tip and Green-veined White butterflies were visiting the self-heal and cuckoo-flowers. In one field margin, we found a spectacular early purple orchid sprouting from the cracking, bare earth. Peering over a luxuriant hedge into a beautiful flowery meadow, we met the disdainful gaze of a Fox slinking through the long grass. We sat down for a break in a similarly flowery meadow, and were rewarded with half a dozen Dingy Skippers buzzing purposefully from flower to flower. Tiny and mottled grey, these moth-like butterflies reward close examination. Those grey wings bear a delicate marbled pattern, and they have charming fluffy faces, with wide-set dark eyes and neat hook-tipped antennae.

Carol, Michèle's mum, met us in Bidborough and joined us for the return journey. Inevitably, she, rather than I, spotted another damselfly, once we were back on lower and wetter ground, though I don't think I would have missed this one as it performed a couple of showy wing-flicks as it settled at about hip-height. It was a male Banded Demoiselle, one of our most striking damselflies. Each wing has a big, glossy dark smudge in its centre. These form bands across both wings on each side when they are open, in flight or in the flicking display employed to entice females, but it looked like today's show was over, as the damsel hunkered down against the breeze and tolerated our close inspection. It seemed to have been spray-painted the most ostentatious metallic violet-blue colour, and then polished to within an inch of its life.

This damsel and its relative, the Beautiful Demoiselle, are the largest damselfly species in Britain. They are at least as long-bodied as the smaller dragons, but have the same slender proportions as other damsels. Banded Demoiselles are common and eye-catching – on still summer days you could see dozens of males along riverbanks, all posing and posturing at the females.

The next day, Rob and I went to Woods Mill, headquarters of the Sussex Wildlife Trust. It was a long way to go to visit a small, albeit lovely, nature reserve, but it fitted in with our plan to visit friends in Brighton, and I knew there was a chance of encountering my favourite of all the Odonata and one that was pivotal to establishing my interest in them – the Beautiful Demoiselle.

There were seven of us – Rob and I, our Brighton friends Mike and Kathy, their eight-year-old daughter Florence and two of Flo's school friends. Mike, Rob and I had brought cameras and big lenses, and crept around with an air of purposeful stealth, trying to catch a glimpse of one of the Nightingales that were singing lustily from the hedges. The children, strangers to the concept of stealth, raced along the boardwalks, leapt across streams, fed the fish in the lake and stuck bushels of goose-grass to Kathy's back. As befits its flagship status, this reserve has something for everyone. The trails pass through patches of damp woodland, skirt around reed beds and wind through meadowland. The centrepiece is a deliciously pretty lake, covered with water lilies and sheltering a sizeable population of fat, lazy Carp that swim to the margins when you stop, hoping you have come to feed them.

We followed a trail past a dry streambed. The banks were thick with vegetation, and it seemed that on every second nettle there was a Harlequin Ladybird, each one different from the last. Some were glossy black with two or four big red blobs, others were orange with a liberal scattering of much smaller black spots, and the variations on these themes seemed endless. Birdsong surrounded us. Mike, freshly returned from a trip to Northumberland, was hearing many of the songsters for the first time this year, and kept stopping us to say 'Whitethroat!', 'Blackcap!', 'Reed Warbler!', and so on.

We all stopped to listen to the first Nightingale. Here is a bird that really does live up to its hype. The song is rich, luscious and full of throbbing passion — the auditory equivalent of a calorie-laden but irresistible chocolate fudge cake. The quality of sound is such that it seems to come with its own acoustically optimised auditorium. As if believing that its physical appearance falls embarrassingly short of the beauty of its voice, the Nightingale usually delivers its masterpiece from deep cover, and try as we might we could not catch a glimpse of it, nor any of the other Nightingales we heard later.

I have in the past had some success at luring Nightingales into view by imitating their songs. Most of the song phrases are much too complex for even an accomplished whistler like myself to mimic, but sooner or later every Nightingale incorporates into its performance a series of long, high, single-note whistles, which might be written down in a field guide as 'seeee, seeee, seeee'. They have a rather strained and breathy quality but are easy to copy and a good rendition *may* cause the Nightingale to poke its head out of its hiding place to check you out. The reward, a three-second glimpse of a featureless, feathery brown face, probably doesn't justify the stress caused to the Nightingale in the brief moments that it thinks its territory has been invaded by an exceptionally unmusical rival Nightingale.

We reached the point on the trail where Rob and I had seen Beautiful Demoiselles last year. There was nothing. It was a little early in the year, and a little breezy too. I was also pretty sure that when we'd been here last year, the dry streambed had contained a stream. Was this a consequence of the almost non-existent April rainfall of 2011? Or maybe the newly dug artificial stream winding across the field below us had something to do with it. We lingered a little while, noting our first Swift of the year zipping overhead, and then finally I did see a flicker of dark damselfly wings at the top of a small tree — a male Beautiful Demoiselle. It settled almost out of view, and just as I was lining up a photo of the tiny part of it that I could see, it took off again and flew up and out of view into the foliage of the much bigger next-door tree, not to be seen again. So while I could add it to our list, I was disappointed, and already thinking about a return trip to give this wonderful damsel the attention it deserves.

The day had one last Odonata treat in store. We were walking towards the big lake that is the centrepiece of this reserve when Rob spotted a vivid yellow dragon squatting quietly on a dry upright stem. It was a Broad-bodied Chaser, even fresher than the one Susan and I had found. Its wings were fully expanded and spread, but its hefty abdomen described an S-shape and looked a little under-inflated. In fact, it looked awfully vulnerable, sitting in full view on its perch. Perhaps the superficial resemblance to a Hornet would prove its salvation, and discourage any passing birds from picking it off. We all gathered to admire the dragon, but its magic didn't enrapture the children for very long and they were soon racing away to the next excitement — another view of that lovely lake and its army of gigantic Carp. Once again the big fish swam to the shore to greet us, their wide mouths gulping in anticipation of more chunks of bread.

The Broad-bodied Chaser had saved the day somewhat, but I knew I wasn't going to settle for such an unsatisfactory view of a Beautiful Demoiselle. Too many years had passed since I last spent quality time with my favourite Odonata species, and so the first thing I did when we got home was start to hatch a plan to put that right.

CHAPTER 7

Calopteryx virgo
The Beautiful Demoiselle

❧ ℥ ❧

After that disappointingly brief glimpse of the year's first Beautiful Demoiselle at Woods Mill, I was keen to arrange another meeting with this lovely insect. We could have returned there, but I wondered whether there might be a place closer to home.

Beautiful Demoiselles aren't exceptionally common in the UK but we lived in the middle of one of the parts of the country where they are 'locally abundant', not far, in fact, from where I'd seen my very first one several years ago. The map in the field guide showed that our home town Sevenoaks fell in the middle of a blob of dark green, indicating a population stronghold for the species — we should have been surrounded by them. Yet I didn't know a local site. However, a little research soon helped me out.

Since late 2009 I've kept a wildlife-watching blog, documenting sightings both locally and further afield. Gradually I've added other similar blogs to my reading list, and it was on one of these that I found what I was looking for. My fellow blogger Warren has a local patch near Tonbridge, just a few miles from Sevenoaks, that he visits virtually daily and blogs about just as frequently — a most impressive undertaking. Putting in the hours brings the rewards, and Warren has recorded a fabulous range of species on his unexceptional-looking patch of farmland, woodland and streamside. Although mainly interested in birds he documents other groups too, including insect life.

Browsing Warren's blog, I was excited to find that he regularly saw Beautiful Demoiselles on his patch — there were even some lovely photos to prove it. I fired off an email asking for more information and soon got a reply, advising me to try the grounds of Hadlow Agricultural College, where I would surely find my quarry.

As luck would have it, I was about to celebrate my birthday and my friend Susan had offered to drive me to a wildlife-rich place of my choice on the big day for a walk and pub lunch, as a present. The forecast on 11 May began well, with warm sunshine until about pub lunchtime whereupon the rain would arrive. As Susan is not one of the earliest risers, I picked nearby Hadlow College, figuring that even if we didn't get going until 10am we would have a couple of hours of productive damsel-hunting time at least.

The college is set in an expanse of lovely gardens, designed and maintained by the students and open free to visitors. A keen gardener, Susan was in her element, wandering down sunny rose-lined avenues, peering into little container ponds and exclaiming excitedly over the many bizarre abstract takes on garden benches that were scattered around the place. A large and noisy rookery in tall trees adjacent to the garden diverted me. For some reason, I had never managed to take good photographs of Rooks, so tried my luck on them today as they flew to and fro from their nests, cawing desolately. The results were disappointing, mainly because of the birds' annoying habit of placing themselves between the sun and my camera. Other, smaller birds were active among the gardens. A Goldfinch sang from the top of a severe, square-cut hedge, and a confiding Chiffchaff busily searched the twigs and leaves of an elegant little ornamental tree for insects, its plumage as freshly green as the tree's foliage.

We strolled around at leisure and eventually found our way to a pond, visible through the shade of tall willow trees. Although the pathway to the edge of the pond was blocked, as new landscaping was in progress, it was while I was looking around for another route that I noticed a tell-tale dark flicker around some huge-leaved ornamental plant tucked in a quiet corner.

I'm no expert on plants, especially non-native, garden species. However, this one was highly distinctive and proved very easy to find when I looked online later. I Googled 'plants with massive leaves' and found a suspect almost at once. Further investigation confirmed it — the plant I'd found, *Gunnera manicata* aka giant rhubarb, was present in the gardens of Hadlow College, near the pond. I also learned that the plant is native to Brazil, and that its great round, ragged leaves can measure well over a metre across.

What *Gunnera manicata* has that appeals to Beautiful Demoiselles is anyone's guess, but as I approached the monster plant I could see several of the dazzling damselflies flying around it, settling on high vantage points to flick their wings before darting out again to chase down some tiny midge or other prey. There were males and females, all fully mature and coloured up, and they made the most wonderful spectacle.

I selected a perched male and started to take his photograph. He was sitting daintily along the edge of a high leaf, in typical down-tilted demoiselle stance with his rear end aimed skywards and his body at 45 degrees to the horizontal perch. In full-on sunlight, his colours were astounding, almost preposterous. The body, from head to abdomen tip, was a rich, deep, iridescent blue-green. Against this, the broad, deep-based wings were a contrasting colour that somehow combined shades of copper and violet, becoming bright orange at the bases where they rooted into the thorax. The creature balanced on splayed out, spindly black legs, and regarded the world with deep brown eyes.

Close to this male sat a female. Her colour scheme was slightly subtler, but not much. The head, thorax and abdomen were bright, deep emerald green, and her glistening wings were golden. This jewel of a creature, not content with simply being beautiful, was also busily showing off by rapidly flicking her wings open and closed. I was puzzled by this behaviour, which

I had seen performed by male Banded Demoiselles and had assumed was part of a courtship display, to attract females. So was this individual actually an immature male? I thought not. Her eyes were dark, not glassy pale, suggesting she was fully mature. If her wing-flicks were intended to attract a male, it wasn't working, none of the nearby males showed any interest. She jumped off her perch to snag a passing fly and returned to the same spot to eat it, her complicated mouth moving in several directions at once as she consumed the poor little thing piece by piece, with great delicacy. Then the wing flicking started again, and after many failed attempts I managed to photograph the moment of full spread, the white pseudopterostigma visible on each of her four wings.

This most glorious of damselflies tends to attract attention wherever it goes. Even the noted 18th century naturalist Gilbert White, best known for his ground-breaking observations of birds, noticed the demoiselles in early summer and made brief mention of them in his writings about wildlife around his home in Selborne, Hampshire. He used the old name of 'Demoiselle Agrion', the word 'agrion' meaning 'wild'. In fact, *Agrion* was formerly the genus name for the demoiselles, before being changed to the rather more descriptive *Calopteryx* (meaning 'beautiful wing'). The Beautiful Demoiselle's species name, *virgo*, means 'virgin' of course (or 'unmarried woman' if you wish to be very coy). 'Demoiselle' is just the French version of 'damsel', so the English name means a 'beautiful girl' or, if you call it a Beautiful Demoiselle Damselfly as some people do, a 'beautiful girl-girl-fly'. I wondered how a male Beautiful Demoiselle would feel, if he were able to consider it, about having such a feminine epithet. But to be fair, the majority of Odonata names, both English and scientific, highlight some physical trait seen in males but not females.

At this point, Susan found me and I cheerfully pointed out the demoiselles on view. She was gratifyingly impressed, and tried to coax the nearest to climb onto her hand. It was having none of this and departed with elegant, bounding flight. Then a shadow crossed the sun, and somehow all the other damsels vanished away as well.

We retraced our steps, and after a little more wandering found our way to the shore of another, larger pond. This one looked quite newly dug, with little in the way of vegetation around its muddy margins. The cloud was closing in and a chill breeze had picked up — rain was not far away. I had a good look in the shoreline vegetation and was rewarded by the discovery of a pair of Azure Damselflies quietly mating on a gracefully curved sedge leaf. The male sat on the top edge of the leaf, the female on the downward curve. She was hanging onto the leaf with her four back legs, which were positioned astride her own swept-forward abdomen, and she had a good grip on the male's abdomen with her front legs.

Looking at the two of them, their long, jointed bodies bent almost double to create this bizarre and inexplicable configuration of copulation, I felt that they provided a really rather apt symbolism for love — the extent of compromise and co-operation required, the vulnerability of their attachment to each other, and the happy coincidence that together they formed a pretty good approximation to a heart shape. They would remain like this for half an hour or more, a significant chunk of their short adult lifespan, and the male would keep hold of his partner for the next hour or so as she laid her eggs. But for damselflies, monogamy is not a satisfactory way of life. A long-lived Azure of either sex can mate a dozen or more times before it dies, and parent a crop of more than 4,000 eggs.

CHAPTER 8

Ischnura elegans
The Blue-tailed Damselfly

❧ ☙

As May 2011 progressed, more and more damselfly nymphs came clambering out of their lakes and ponds into the fresh air, ready to transform into their winged forms. I spent a few sunny May mornings at our most local reserve, where I knew I would see the commoner species.

Sevenoaks Wildlife Reserve was the very first UK nature reserve to be founded around a series of ex-gravel pits. These days, this is common-place. Extract the gravel, allow the resultant cavities in the ground to flood, encourage wetland vegetation to colonise the margins, declare the whole thing a nature reserve and manage it accordingly. Most counties have at least one nature reserve that used to be a gravel extraction site. None are as venerable as Sevenoaks Wildlife Reserve. Originally managed privately by the Harrison family of Sevenoaks, the site is now under the care of Kent Wildlife Trust. It lies along the river Darent, which separates the town of Sevenoaks to the south from the sharply rising hills of the North Downs.

It takes me 20 minutes to walk here from home. Once I've made it across the busy A25 I head down the access track past a riding school where I'll sometimes see the horses and ponies being led in slow circles around their big field, each with a nervous child perched on top. Most times I'm too early for this, and the horses are standing around staring into space or cropping the already well-cropped grass. Sometimes I'll disturb a Magpie that had been sitting on a horse's back, pulling hair from the poor ani-mal's mane to use as nest lining.

Alongside the reserve's car park is a 'nature garden', planted with a mix of wild and cultivated flowers, with a shallow little pond as its centre-piece. In here, I'll often see the first damselfly of the day.

This time, in the garden, I spot a damsel resting on a flower stem, its body sticking out at a right angle, and lean in for a closer look. There's still a bit too much early morning chill for flight to be a favourable option, but the damsel is not keen on being scrutinised so it shimmies around to the far side of the stem. However, because its head is wider than the flower stem, I can still see its spherical eyes, one on either side. They are pale with blurry dark spots in the centres — the 'pseudo-pupils' that are the points of highest visual acuity. If I move my head left or right the damsel shifts its position accordingly, keeping the stem between us, like the Pink Panther using a lamppost to hide from an enemy. To identify it, because I'm still not very clued-up on the damsels, I back off and take a photo to examine later.

I head past the biggest lake on the site. It has a cluster of low islands, which at this time of year are home to a few pairs of Lapwings and Little Ringed Plovers. Both are fiercely defensive of their chosen patches, the males taking off every couple of minutes to harass a passing crow, heron or even a tiny Pied Wagtail. I get glimpses of the water as I walk up the trail through a hand-planted woodland. Blackcaps that have been singing here since April have suddenly been joined by a choir of Garden Warblers — the latter giving a quieter and more hurried song in contrast to the Blackcaps' loud and fluted outbursts.

Through the wood runs the river Darent, clear and shallow. Where it passes a thick bank of nettles, I find my first Blue-tailed Damselflies of the day. In flight, these small damsels are improbable-looking creatures. The blue on their thorax and the rather bulbous abdomen-tip is eye-catching, but the main length of their bodies is black and narrows to an incredible thinness. On the move, in certain lights this black thread of body some-times virtually disappears and only the blue bits are obvious, so what you see is two beads of blue that don't look connected to each other, but move around as one.

Male Blue-tails have this simple colour scheme as do some females, but others are quite different. One resting female I see has a beautiful rose-pink thorax, the colour spreading onto the base of the abdomen. This is the form 'rufescens' — perhaps the prettiest of the five or so distinct colour forms you may see among female Blue-tails. According to my field guide, the 'rufescens' females are youngsters, destined to fade into the more drab fully mature form 'rufescens-obsoleta'. Unlike many damselflies, the young or teneral adult Blue-tails of both sexes stay close to the water, rather than moving away to drier spots over the few days it takes them to reach full maturity.

There is much speculation over why female Blue-tails are so variable. Immature ones may be either 'rufescens' or 'violacea', the latter also very beautiful with a violet-blue thorax. While the 'rufescens' will become 'rufescens-obsoleta', the 'violacea' females can go one of two ways — either they will become the drab greenish 'infuscens', or the andromorph or male-like form, which is coloured just like a mature male. Now, male Blue-tails find potential mates mainly by sight, so where (and indeed when) do these cross-dressing females get off? Most males will overlook them, so their chances of successfully breeding are reduced.

On the other hand, the more feminine females don't necessarily have things particularly easy. They have plenty of mating opportunities, but are more likely to suffer injury or even premature death when caught up in a scrum of eager males. Perhaps the male-like andromorph females lead longer and less harassed lives than the 'infuscens' and 'rufescens-obsoleta' females because they receive less attention from the real males, although they would still need to mate at some point. It seems logical that in years with an overabundance of males the cross-dressers will do better, picking up mating opportunities with males that failed to get lucky with the conventional females. Conversely, years with fewer males might work out in favour of the girly girls. Whatever the answer, there must be long-term advantages to both strategies, or they would not persist.

All these italicised names are alienating to some, but I like them a great deal, finding them rather more poetic than their English translations. Both *rufescens* and *violacea* are words to describe colour — rufous and violet respectively. So is *infuscans* — that means 'tinged with brown'. The Blue-tailed's genus name, *Ischnura*, is one of my favourite Odonata words, though all it means is 'slim tail'. The second part of the name, *elegans*, means exactly what it sounds like. So an *infuscans* form Blue-tailed Damselfly, for example, is an 'elegant slim-tail with a tinge of brown'.

I carry on along the trail. At the top end of the track is the small Snipe Bog Lake. Here I hang a right and head to a sort of sunken marsh area that is really good for damselflies. Things are warming up and the insects are starting to fly. The teneral damsels, those that have only recently emerged, are the most eyecatching at a distance because the light catches their very shiny wings, but without their full adult colours they are confusingly hard to identify. The mature male damsels have bright blue bodies, but their wings have lost their shine. In flight, the wings are almost invisible, making the insects look like animated blue needles, bobbing slowly and erratically around the reeds and sedges at the edge of the marsh.

There are two species of mainly blue damselflies that are very common throughout most of the UK. They are the Common Blue and the Azure, and at first glance they look exactly the same. You need a close look to see the distinguishing features, but the damsels may not tolerate this. I usually take photos so I can check at leisure, as damsels don't always allow you to peer at them at point-blank range, but I find a surprising alternative way to study a damsel at close quarters when I notice one caught by its wing in a spider's web.

Spiders don't normally hang around when something this big gets itself caught in their traps. The spider must have been out, or otherwise engaged. I quickly and carefully extract the damsel and settle it onto my fingertip, where it sits quietly. I wonder how long it had struggled in the web. A piece of silk trails from its wingtip — I gently pull it away but there is some more that I can't get at without risking damage to the little creature's wings.

At point-blank range, I can see that this male damselfly is a Common Blue. The second segment of his abdomen, close to the thorax, is blue, with a black marking that looks like a rugby ball on a little stalk. If this had been an Azure, the marking on this segment would be a delicate U shape. The Azure also has a little black half-stripe (the 'Coenagrion spur') on its thorax-side that the Common Blue lacks. These two common damsels aren't actually that closely related. There are several other species in the genus *Coenagrion* in Britain alongside the Azure — the Northern, Southern, Irish and Dainty are all *Coenagrion* species — but the Common Blue is our only representative of its genus — *Enallagma*.

Taking a wider view, the two genera are almost completely separated by the Atlantic. The Common Blue, despite its all-encompassing UK and European distribution, is the outsider here — it is one of only four *Enallagma* species (from a total of nearly 40) that's found in Europe, with all the rest in North and South America (where they are called 'bluets'). The situation is reversed — and even more marked — when we look at *Coenagrion*; of about 50 species, only three occur in America This pattern of distribution makes it pretty clear where the two groups originally evolved. *Enallagma* species radiated out in all directions from the group's 'centre of diversity' in eastern North America. The ancestors of the Common Blue Damselfly went east, and crossed from America to Europe many millennia ago, when the two continents were joined by land. Those filmy little wings aren't built to cross 6,000 kilometres of ocean.

For those with an interest in etymology as well as entomology, the Azure's genus name *Coenagrion* was coined when it was decided that it and related damselflies belonged to a separate genus to the demoiselle damselflies — previously, all were jumbled together in the genus *Agrion*. The 'coen' part of *Coenagrion* means 'shared', presumably to indicate that these damselflies once shared a genus with the *Agrion* species, the demoiselles (whose genus has now been renamed to *Calopteryx* anyway). The Azure's second name, *puella*, is Latin for 'girl', alluding to the insect's pretty, delicate appearance.

The word *Enallagma* is a combination of two Greek words meaning 'alternate splinters', which may refer to the alternate blue and black stripes on the damselfly's body. The Common Blue's species name, *cyathigerum*, has a much more clear-cut and pleasing translation, although purists won't be impressed to learn that it is derived from two different languages. It is a combination of the Greek word for 'cup' (*kyathos*), and the Latin word for 'carrier' (*gerula*). So the name means 'cup-carrier' and refers to the shape of the marking on the male's second abdominal segment. To me it looks like a ball on a stick but it's easy to see how it can also look like a slim-stemmed cup or glass.

Ancestrally speaking it may be an American abroad, but this Common Blue seems comfortable enough on my hand. He recovers some energy and begins to walk clumsily about, his long body wobbling from side to side. Damselfly legs and feet are better at clinging and grabbing than they are at walking, but the hooked little claws that tip each leg give him a stable grip and ensure he doesn't fall over — I can't feel a thing though, so light is the touch. After a bit of exploration, the damsel stops, and begins vigorously bending and flexing his abdomen.

The damselfly abdomen is made up of 10 tubular segments, and the seams between them are not obvious unless you look closely. Yet the abdomen is as flexible as a cat's tail, at least in the up-and-down plane, and my damsel is curling and straightening his body at high speed, almost flicking it back and forth. After a while I work out what he is doing — he's trying to clear out the last remaining bits of spider silk that are caught on his wings, using his body as a sort of brush. Again and again he passes his abdomen between his wings, then pauses to test the wings with a quick little buzz, then back to the cleaning ritual. Finally the wings are clean, and the damsel lifts off neatly and flies down towards the water's edge.

I carry on with my walk. From the reeds I can hear the tireless squeaky song of a well-hidden Reed Warbler, and the simple tweet-tweet-rattle refrain of a Reed Bunting from the top of a young willow tree. A small squadron of Sand Martins chases overhead, snapping up the aerial 'plankton' of tiny flies and baby spiders on their silk parachutes, swept high over the trees on the warm rising air. I follow the trail to Long Lake, prettiest of the former gravel pits, and the place where I'll hopefully find a fourth damselfly species.

One of the best bits of kit I've ever bought for wildlife watching is a green rucksack that turns into a simple folding chair. I unfold my chair at the western corner of Long Lake and just watch. There is a small inlet in front of me, and the vegetation on either side of it is thronged with damsels, many of the mature ones paired up. Most are Azures and Common Blues, but among them are good numbers of Red-eyed Damselflies.

At first glance, the Red-eyed Damselfly male looks like the Blue-tailed. It has a light blue thorax and matching abdomen-tip, and is otherwise dark. However, it is a bigger and more robust creature, and while the Blue-tailed has blue and black eyes, those of the Red-eyed are ... well. In good light they glow a deep and vivid scarlet. If the light's is poor they just look dark, but they always look strikingly one-toned and, because of this, particularly bulbous. Female Red-eyes have the same pop-eyed look, and colourwise are blackish on their uppersides and apple green underneath. The scientific name of the Red-eyed Damselfly's genus, *Erythromma*, means 'blood-red eyes', although at least one other species in the genus actually has bright blue eyes. Its species name, *najas*, comes from 'naiad', a Greek water nymph.

Red-eyed Damselflies do stick closer to the water than many of the other damselflies, which may wander away from water some distance to feed. They like resting on lily pads, of which there are plenty on Long Lake. Sure enough, the lily pad closest to me has a pair of Red-eyes joined one in front of the other in 'tandem', the male at the front and gripping the back of the female's head with the clasping appendages at his rear end. Resting on the same leaf and facing the couple are three unpaired male Red-eyes, eyeballing the lucky paired-up male and his partner, perhaps looking for seduction tips if not a chance at stealing away the female.

Looking around, I spot another Red-eye couple, on a sedge stem nearby. The male is clinging to the stem, the female dangling underneath with her legs waving in the air, held fast by his claspers gripping the back of her head. Laboriously she curls her abdomen up and forward, under her thorax, to bring its tip into contact with the underside of the male's body, just behind his thorax. Joined at these two points, their bodies gracefully curved into position, the damsels form what is commonly called a 'wheel', although in truth the shape they make is more like a heart. I'm inordinately pleased with the photos I catch of them in position.

The whole process seems overcomplicated and looks more than a little uncomfortable. It's not something you'll see among other insects, which simply join their rear ends together. Sometimes the male gets on the female's back, gripping her in a similar way to how a frog or toad clasps its partner. In large-winged insects like butterflies, where this wouldn't really work, the two join up back to back. Only the dragons and damsels do the head clasping and form the elaborate 'wheel'.

This stage is essentially copulation, as the female is receiving sperm from the male, but the sperm didn't originate up there, it is formed in his 'primary genitalia' close to his abdomen tip, and then he transfers it up to his 'secondary genitalia' at the abdomen base for her to collect with the appropriate bit of her own reproductive equipment — a receptacle called the 'spermatheca'. However, before she can have it, a special organ that's part of the male's 'secondary genitalia scrapes out the female's spermatheca, removing any deposits that may have been left there by his mate's preceding partners.

This last behaviour, whereby the males try to get rid of previous males' sperm, is known as 'sperm competition'. It is practised in various ways in many different animals, and is well developed in Odonata, though the hawker dragonflies' methods are less sophisticated than those of the damselflies. It is thought that sperm competition is what led to the evolution of the Odonata's convoluted method of copulation.

If these had been dragonflies, then once the deed was done, the male would let his partner go and she'd fly off alone to develop and lay her eggs. But male damsels, having gone to considerable trouble to make sure that only their own sperm remains in the female's spermatheca, are determined to protect their investment and see things through, so they keep holding on and the pair returns to the tandem position. When the female is ready to lay eggs, not long after copulation, he is still attached to her and they fly together out over the water. So the female has to lower her rear into the water with the male still attached to her head by his rear end, his long body comically sticking up out of the water. I manage a photo of a Red-eyed pair flying together in tandem, and another of four pairs of Azure Damselflies in tandem, the females crouched on floating vegetation ready to lay eggs, and their attached mates poking straight upwards like peculiar little blue water-plants, with wings held swept back and feet clasping at thin air.

As I'm watching the masses of damsels going about this rigmarole of courtship, mate-seizing, wheel-forming and egg-laying, a dragon flies low and fast into the little inlet, hangs motionless in the air there for a beat of two, then whizzes off out of view behind the screening bank of shoreline plant life. Scrutinising its fast-fading after-image in my mind's eye, I see a smallish, dark bronzy-green dragonfly with a slim body that swells to a more bulgy, almost hooked tip. It's a species I've never seen before, but I think I know what it is.

Here in the UK there are three species of 'emerald dragonflies' from the family Corduliidae. This is one of them — the Downy Emerald. I hope fervently that this two-second view is just the start of my acquaintance with this new (to me) dragon, and my wishes are fulfilled — a few minutes later the dragon is back. Again it pauses before darting away. Its look is very 'dragony' — the glossy greenness, the clear-cut angular shape of its body — but its marvellous precision flying gives it an aura that's less mythical beast and more futuristic miniature spaceship. It has the most vivid green eyes, with a scattering of dark spots, while the sheen of its body is slightly dulled by its coat of fine fuzz. When it pauses in flight it hangs so steadily that I think getting an in-flight photograph will be a breeze.

Half an hour later, the Downy Emerald has been patrolling the lake-shore non-stop. Every couple of minutes it drops into the inlet, hovers neatly on the spot for precisely as long as it takes me to lift up my camera and get it in the viewfinder, then zips off before the autofocus locks onto it. I have a couple of frames showing a greenish blur against a sharp back-drop of undisturbed water, and that's all. I decide to try manual focus, and capture a few more frames of blur, as the dragon moves too fast for my hand on the focusing ring. It's an hour before I manage to fluke a few sharp photos. Relieved, I put the camera down and concentrate on just watching.

I know my Downy Emerald is a male. Females have a more parallel-sided body shape — unlike humans, in dragons it is usually the male who has the hourglass figure with a pronounced 'waist' and proportionately bigger backside. The sexes also show quite different behaviour. Mature males set up a territory along the shore of a suitable breeding lake, and patrol it for as long as the sun shines. They will stop at suitable spots, tilt-ing their bottoms skywards to better exhibit their glossy green abdomens, and at the territory boundaries disputes occur. Females hunt away from the lakes and only come to the water when ready to breed.

The Downy Emerald is the only British species in the genus *Cordulia* (the other two emerald dragons that live here look very similar to it but are classed in a separate genus). While *Cordulia* sounds like a variant on the name 'Cordelia', your baby daughter would probably prefer you stick to the latter when considering names. While 'Cordelia' means 'cordial of temperament', *Cordulia* is Greek for 'club-shaped', not a look any woman would really aspire to but a perfect description of the Downy Emerald's silhouette. Its species name, *aenea*, is similarly functional, meaning 'bronze-coloured' in Latin.

The sun goes in for a little while, and sure enough the Downy disappears. I carry on walking, watching clouds of teneral damsels lift off from the grass at my feet, then I lurch to a halt just too late to avoid disturbing the Downy Emerald that was resting on a bramble clump just in front of me. It flies quickly up into a treetop. I'm cross with myself but there's nothing to be done. I stop at another inlet, the sun comes out again and there are soon Downy Emeralds patrolling the lake shore again. This time I think I'm close to a territory boundary, as I see a few mid-air skirmishes. I also see damselflies rushing at the Emeralds, as if trying to drive them away. Small birds frequently mob predators, but I've never seen small insects doing it before. It's a risky business, to say the least, but I don't see the dragons trying to turn the tables on their attackers this time. Far more dangerous are the many spider webs strung among the plants by the water, and I rescue a couple more trapped damsels. I'm interfering with nature and stealing a spider's hard-earned lunch, but I can't resist the chance to hold the beautiful little insects on my hand for the few moments it takes them to recover their composure and fly away.

I take a different route back to my starting point, passing a lively stretch of the River Darent as it bubbles through a corridor of fast-growing nettles. Here, as I'd hoped, I find a few Banded Demoiselles. They, too, are in courtship mode. Like their relatives, the Beautiful Demoiselles, they are more sophisticated about the whole process than the small damselflies, with the males performing an obvious and rather beautiful display to arouse the females' interest. This involves much flicking of the wings, showing off the dark blue band that marks the centre of each. The males settle on a suitably prominent leaf that shows them off to good advantage and flick their wings at the females. Males that decide they are too close to each other fly up into the air together and squabble gracefully in mid-air before the victor returns to his preferred spot, and the vanquished finds a new stage.

Back at the car park, I'm putting away my camera gear when I spot a fat, chunky dragon gliding around the trees a few metres up. I watch as it circles and then settles neatly at the tip of a bare twig. In the sidelong sunlight it glows a rich, 'hornetty' yellow. It is a Broad-bodied Chaser, its brightness suggesting it is a young individual. I unpack all my stuff again and go over to where it's settled — too hastily, as I disturb it and off it flies. But a chaser doesn't give up its favourite perch that easily. Soon the dragon is back at exactly the same spot, settling neatly with head upwards and its broad slab of an abdomen pointed down at me. Its wings, longer and wider than its body, are marked at their bases with a triangular suffusion of brownish-black.

The differences in shape between forewings and hindwings is very obvious as I look up at the chaser — the forewing parallel-edged with a little upward kink at the 'elbow', the hindwing broad at its base and tapering to a narrower tip. This mismatch between the wing shapes is evident to a greater or lesser extent in all dragonflies, and is one of the key differences between dragons and damsels, the latter having much more closely matched wings.

These are the typical late-spring species on my local patch, although until this year I hadn't known there were Downy Emeralds here. On one of my May visits I added one more species, although June is a more typical month for the species. It was a Black-tailed Skimmer, fresh from the water and resplendent in its adolescent uniform of black and lemon yellow — if it was a male it would gradually turn light blue over the next week or so.

I photographed it as it rested on the wooden slats marking out a fishing swim by the big lake. Later in the season I'd see many more Black-tails here, along with a whole new set of summer species, but for now there were other places to go and other Odonata to see.

CHAPTER 9

Coenagrion pulchellum
The Variable Damselfly

⚜ ⚜

Every year in early January, I get together with an assortment of former colleagues and sometimes some extra friends, for a day of birdwatching. We meet early and follow a tried and tested itinerary, working our way along the East Sussex and Kent coast, pausing for fish and chips at Dungeness, and then heading back to finish at Rye Harbour at dusk, where we'll watch the Bitterns flying in to their reed bed roost at Castle Farm. It's a great day out, even when the weather conspires against us. Nigel has been keeping track of everything seen since he kicked off the tradition some 25 years ago and the cumulative bird list has topped 150 species, although we've not yet quite made it to 100 on a single day.

That's the winter 'Big Day', but in some years we also manage a spring Big Day. The goal is the same, to see as many bird species as we can in one day, but because there are so many more hours of daylight in May than in January, the spring Big Day takes a bit more energy and commitment. After weeks of emailing back and forth, a hard core of four of us (Nigel, Jim, Rob and I) fixed a date and I hatched a plan. While I knew the day's focus would very much be on birding, I had hopes of finding the odd Odonata among the warblers, waders and wildfowl.

This time we had decided to confine our activities to Kent. As the only one of the three keen birders who actually lives in Kent, it fell to me to devise an itinerary. I decreed that we should meet at Bough Beech Reservoir at 8am, just a 15-minute drive from where Rob and I lived and short enough that I could make us a hearty breakfast of bacon rolls that would still be hot by the time we met.

Rob and I drove through beautiful, intensely green Wealden country-side to reach the reservoir, traversing Ide Hill, Kent's highest point. From this part of the road you can see a wonderful spread of woods, farmland and expensive little country houses, while the reservoir itself glints away in the middle distance. Then the road dives steeply through woodland, where a small herd of Fallow Deer edged away from the roadside as we went by, flicking their long tails.

A sharp left took us past the Bough Beech visitor centre, out to the narrow causeway road that cuts off one corner of the large, teardrop-shaped reservoir. The causeway is a traditional gathering place for birders. On the occasions that a really rare bird is found the road may be lined for the best part of a mile with vehicles and crowds bearing expensive optics. I'd only seen this once, when a Little Crake showed up in 1997, but the reservoir has pulled in other exciting rarities including a Spotted Sandpiper from North America and a Radde's Warbler from East Asia.

This morning, only one other car was present – Nigel's. The man himself was busy looking for birds, tripod and telescope set up and pointing out at the water. His happy smile intensified considerably when I handed him a bacon roll. After we'd breakfasted, enjoying views of Common Terns gracefully skimming over the reservoir as we ate, Nigel explained that Jim had decided to meet us later at Elmley Marshes on Sheppey, as the roads to Bough Beech are fiddly and complex, and he had doubts over his navigational skills. This left us with a spare breakfast, which after a mini-mum of soul-searching we ate, on the spurious grounds that Jim surely wouldn't want a cold bacon roll.

Using Nigel's telescope we scoured the water for one of the local specialities (Mandarin Duck) to no avail, and listened at a nearby copse for another (Nightingale) with similar results. We walked on to the visitor centre, where there is a bird feeding station set in an apple orchard, and checked the feeders for a late Brambling – these handsome winter-visiting finches sometimes linger into mid-spring. But again, we had no luck. On the return walk we did hear a few hesitant bursts of song from a well-hidden Nightingale, so all was not lost. Even though the bird was not in full song, the richly throbbing, voluptuous quality of the few fluted notes we heard was unmistakeable, and quite enchanting on that still, misty morning.

We paid a flying visit to my local patch, Sevenoaks Wildlife Reserve, after that. No time to look for Odonata, we were there and gone within half an hour, having seen Egyptian Goose and Little Ringed Plover. The Egyptian Geese (there were four of them) were helpfully sitting in the large field by the access track. An introduced species, it is a striking bird, decked out in shades of tan and chestnut with a possibly malevolent gleam in its yellow eyes. A hundred yards from the car park at a viewing mound across East Lake we saw the Little Ringed Plover, a perky little wader with a striped chest, which ran on clockwork legs to and fro on the shoreline of a gravelly island. The morning mist was beginning to melt away, unveiling a beatific blue sky mid-morning as we took the old bridge to Elmley Marshes, crossing the broad grey sweep of the Swale to the Isle of Sheppey.

Had I but known it, we had driven right past the very site where the Dainty Damselfly, a pretty little blue damsel, has staged its comeback as a British breeding species, almost 60 years after becoming extinct in its former stomping grounds in Essex. The site is an unprepossessing pool beneath a flyover, easy enough to get to but so bird-focused were we that even if I'd known the Dainties were there I doubt I could have persuaded the team to make a detour. However, much of the Isle of Sheppey is covered with marshland, creeks and pools, all potentially suitable damselfly habitat, and Elmley Marshes, an RSPB reserve replete with watery habitat, may well be home to Dainties in future.

Jim was waiting for us at the start of the access track, a tall man folded into a tiny car. He greeted us cheerily enough but expressed sorrow and indignation on discovering what had happened to his bacon roll. After appeasing him, we set off slowly down the two miles of access track, for along this rough road through an expanse of grazing marsh there are often lots of birds to be seen.

The wet grassland is home to nesting waders, in particular Lapwings and Redshanks. The former were everywhere, trotting about in the fast-growing grass, their tremendous long crests swishing in the breeze, and many were in the air as well, displaying over their territories with whooping calls and loop-the-loop flight on their piebald, ping-pong bat wings. The Redshanks, brown and speckled, stood tall on glowing scarlet legs and warily watched us crawl past at five miles an hour. There were other birds to see as well — a small crowd of Mediterranean Gulls on the track up ahead, a Skylark standing on a little hillock ready to launch into song, and a male Yellow Wagtail, slim, fast-running and incandescently bright, snapping up the swarms of flies attracted to piles of manure placed alongside the track for that very purpose by the reserve wardens.

The grassland is cut through with narrow, reedy ditches and here there were a few Hairy Dragonflies, just the occasional little flash of blue as a patrolling male moved up over the tops of the reeds for a moment. No doubt there were also a few damsels in those ditches, but there's no getting out of the car to look along this track as there is too great a risk of disturbing the nesting birds.

Elmley is a large and wonderful nature reserve, with pools, scrapes and foreshore to enjoy, but it is also a very long walk around and we had no time for that — we were only here for what we could see along the track. So at the far end we turned around, and then drove back. On the return journey we found a Brown Hare crouched almost out of sight in the long grass, its long lily-petal ears lowered flat against its hunched back. It crunched through a mouthful of greenery and regarded us through half-closed, amber eyes. Brown Hares are rare in Kent generally, but on the Isle of Sheppey there are plenty of them.

Our next site was Oare Marshes near Faversham, back on 'mainland' Kent and a little further east. You pull in off the little dead-end road just before you reach the Swale and there before you is the East Flood — a large, shallow pool with multiple islands, which is magnetically attractive to water birds of all kinds. It is at its best in late summer and autumn, as great flocks of waders from the Arctic — the youngsters of the year and their parents — stop off on their southbound migration.

Out on the water were various ducks and waders, and in the ditch right beside the road another couple of male Hairy Hawkers, gliding to and fro on the lookout for females. What really caught our attention, though, was the Avocet family on a tiny island very close to the road. Mum and dad were supremely elegant, sleek black-and-white supermodels of birds, with slender upswept bills, strutting on stilt-like bluish legs. Their four chicks, by contrast, embodied avian cuteness — little grey fluffballs well short of their parents' ankle height, with stuck-on bills and bright button eyes. They skittered to and fro on their little muddy island, while the parents kept a watchful eye out for danger.

A short time after we'd arrived, potential danger did make an appearance. A Little Egret — pure white, leggy and long-necked, and bedecked with delicate long nuptial plumes on head and back — came flying unconcernedly towards the Avocet family. Little Egrets belong to the heron family, and as such are hunters, taking fish, frogs and yes, perhaps a fluffy chick or two. Whether this egret was intent on attacking the little Avocet family or not was a moot point though, as its approach triggered an immediate pre-emptive response from the parents. The male Avocet flew straight out on an interception course, twisted in mid-air, and gave the startled egret an almighty kick in the head. This knocked it sprawling into the water. The egret struggled to take off again, splashing wildly and squawking loudly, but the female Avocet had joined in now and the two of them rained down blows upon their hapless victim. The egret half ran, half flew along in a great spray of water and managed to lift off properly after a few strokes. The Avocets let it go then and it flapped away with more croaking yells, its lovely white plumes all stuck together and dripping pond water. The Avocet pair rejoined their chicks, with an air of triumph and the satisfied dusting-off of fists. Rob and I caught the whole drama on camera.

Our next port of call was Stodmarsh, a National Nature Reserve in the lovely Stour Valley near Canterbury. The river Stour springs into being some way west of Ashford, and winds its circuitous way west to flow into the sea just south of Ramsgate. The stretch east of Canterbury is associated with some extensive and fairly unspoilt wetlands and reedbeds: besides the 618 acres of Stodmarsh and the adjoining Grove Ferry, there is the 100-acre Westbere Lake and many smaller pools. The area is noted for attracting dozens of Hobbies in mid-spring — these insect-eating falcons arrive just as huge numbers of St Mark's Flies are hatching out. We were a little late in the season for the peak Hobby numbers, but as we walked out across pretty pastures towards the reedbeds at Grove Ferry we soon saw several Hobbies hawking over the fields. A bit of a wind had picked up now and the Hobbies rode the breeze with skill and style, occasionally swinging their legs forward (revealing their russet-red feather 'trousers') to catch a fly.

From deep in the reedbeds came the monotonous, ceaseless songs of Reed Warblers, plus another sound, a dry, tapping, metallic 'ping' which we all knew to be Bearded Tits. These lovely little birds, ginger with long tails and mournful-looking drooping black moustaches in the males, are reedbed specialists, feeding on insects and reed seeds and climbing about in the tall stems with parrot-like dexterity. However, they don't like the wind, and stay down and out of sight if things are too breezy. Today all we had were a few glimpses of one or two in flight — little ball-and-stick shapes bobbing up over the swaying reed heads for a moment then dipping back down.

A look in one of the hides over a shallow flash of water provided us with views of a Greenshank, pale grey and lightweight, stepping daintily through the water on long silver-green legs. As we headed back, we noted a great cluster of spiny black Peacock butterfly caterpillars engulfing several stinging nettle plants. It was shortly after this that I spotted a blue damselfly sitting on a thick reed blade.

I stopped, partly because this was one of only half a dozen Odonata I'd seen all day, and partly because there was something about it that looked different to the Azure and Common Blue Damselflies I'd seen on my local patch. None of the others stopped with me, and I let them go off into the distance while I carefully edged close enough to the quietly resting damsel to take some photos.

I knew that you could tell a male Azure from a male Common Blue in a top-down view by the thickness of the antehumeral stripes – a pair of blue lines running longitudinally down the top of the otherwise black thorax. Broad stripes = Common Blue, narrow ones = Azure. But this damsel had none; his thorax top was entirely black. I moved my gaze downwards to check the second abdominal segment where I would find another identification clue – the shape of the black marking here. A stalked blob = Common Blue, a delicate U-shape = Azure. Again, this damsel matched neither, the marking there was a thick, stalked U-shape. The rest of the abdomen seemed to have more black on it than I recalled being the case with either Azure or Common Blue.

Realising that I had found something that was neither a Common Blue nor an Azure, I made sure I took plenty of pictures of the damsel, which had still not moved a muscle, and then had to go chasing after the others, who were almost back at the road. I hadn't brought my dragonfly field guide with me but the reading that I'd been doing in previous weeks was telling me that I'd just found my first ever Variable Damselfly.

The Variable Damselfly is a much more scarce species than the Azure, except in Ireland where it seems to be the commoner of the two.

In England, it occurs patchily, though one of its patches is east and south Kent. It could well be that I had seen it before back in the days where I really didn't know one blue damselfly from another. A member of the genus *Coenagrion*, its species name is *pulchellum*, which is Latin for 'pretty'. It certainly is pretty, though not really more so than any other damselfly — though when I looked in the field guide later on I did think the illustration of the female 'blue form' was very appealing, with her evenly spaced blue and black bands.

We spent the last hours of the day at Dungeness, chasing after birds of land and sea. Sitting on the shingle beach with our backs to the eyesore of a power station, watching sunlit Sandwich Terns plunge-diving into the unnaturally warm and therefore fish-filled sea by the outflow, we reflected on an excellent day. Jim, with a hot date to attend, bade us farewell at this point but Nigel accompanied Rob and I on a walk around the power station's perimeter fence, where we found a female Black Redstart. This bird, one of my favourite species of all time, is rare in Britain but has an unaccountable attachment to huge, ugly buildings, as well as to urban wastelands in general. The smoky-grey, red-tailed little bird was fossicking about on the ground at the base of the fence, and then flying up to sit on one of the many strands of barbed wire erected to keep trouble-makers well away from the nuclear reactors.

We moved on to finish the day at Dungeness RSPB reserve. From one of the birdwatching hides overlooking the large Burrowes Pit we found a pair of Garganeys, beautiful small dabbling ducks that are scarce summer visitors to Britain. They came paddling along, the male resplendent with long, drooping scapular feathers on his back and flashy white eyebrows, the female more subtly marked but just as pretty in her own way. They parked up right in front of us and gave their plumage a thorough preen before settling down to sleep.

The Garganeys rounded off a day list of birds that didn't quite reach 100, which I suppose was slightly disappointing, but as we packed up to head home I was still thinking about the 'probable' Variable Damselfly. A look at the book when we got home confirmed it — another new Odonata and, given that today had been strictly for the birds, a particularly satisfying bonus.

CHAPTER 10

Platycnemis pennipes
The White-legged Damselfly

❦ ❧

O ne of the trickiest of all dragonflies to see in Britain is the Common Clubtail or Club-tailed Dragonfly. For a start it is quite rare, only breeding in certain stretches of a few rivers. Add to that its rather elusive ways — it is really only likely to be seen when it has just emerged from the water, before it has gained the strength to fly very far — and you'll appreciate that finding one is an exercise in precision timing.

In 2011, our timing was all wrong. The clubtails are among the UK's earliest dragons, making their maiden flights in the second or third week of May, and nearly all the individuals in a population come out within a week of each other. But for one reason or another we didn't manage to head for our nearest clubtail site until the second week of June. Nor was the weather especially encouraging, with fluffy white clouds building and coalescing all through the morning, though there were some sunny spells now and then. We headed out early anyway, still somewhat fired up with optimism after our unexpected success with the Vagrant Emperor at Dungeness, and drove westwards to a quiet and beautiful loop of the Thames near Reading.

The flowery fields that lined this mellow, pretty stretch of riverside were alive with insect life, and taking advantage of this bounty were numerous House Martins and a few Swallows. The martins' backs reflected a violet gloss when the sun caught them, while the Swallows were bluer. Both of these birds catch insects in much the same way that a whale 'catches' plankton — by opening their mouths wide at the appropriate moment. In fact the morass of little flying things high in the air, including all manner of winged insects but also baby spiders 'ballooning' on a parachute of silk, is sometimes referred to as 'aerial plankton'. Once enough has been engulfed, and compressed in the bird's throat to make a sticky ball of insecty goodness, it will be taken back to the nest and delivered to the chicks.

We reached the 'correct' spot for finding the clubtails, and settled down to wait. I had a quick look at the field guide, which explained that when freshly emerged, the dragons don't hang around the riverside long but head away from water as soon as they have gathered enough strength. This apparent aversion to water is an interesting trait in a creature which has lived in the stuff for up to five years, and an illustration of the double life led by all Odonata — perfectly adapted creatures of water and air, but definitely not both at the same time. Newly emerged clubtails aim for open, sunny woodland where they spend at least a week maturing. Then the males return to the riverside and loiter there, waiting to intercept a mature female. So while we were much too late to catch an emergent clubtail, I did hold out hope that we might find a territorial male. An hour later, though, there had been not a sniff of a clubtail, and these hopes were beginning to fade. I was already planning to be here without fail in May 2012.

The day wasn't a complete washout. There were damselflies around, including lots of Red-eyed that were holding social gatherings on top of mats of floating vegetation. The males used perches like these as watchpoints to scan for females, and showed little hostility towards each other — sometimes two males would happily share a lily pad and there would only be a show of stroppiness if one got very close to the other. Once a male had managed to grab himself a mate he would often return to the same lily pad, as if to show off his trophy to the other males. The general demeanour of this species is rather mellow, with less of the bickering and attempts to interfere with coupled-up pairs than is seen in some other damsels.

Red-eyed Damselflies enjoy a very long flight season, first appearing in May but still around into August. This doesn't mean individual damsels live for months on end but that the emergence period is very protracted with new individuals appearing constantly through the summer — the opposite strategy to that employed by the Common Clubtails, which are really on the wing for only a month.

It's easy to see pros and cons to both tactics. By doing everything synchronously, the clubtails have the best possible chance of finding mating partners, and they make the most of the best possible time of year to be feeding up and seeding a new generation, in terms of day length at least. However, they can't be sure that their month of activity will coincide with good weather. Four weeks of below-average temperatures and sunshine can really ruin the whole year for them — they will live a little longer if they spend more time inactive but that can only take them so far. The Red-eyed Damsels can overcome a few bad summer weeks here and there, although their population at any given time is well short of the potential maximum.

Another species active around the riverside was the Banded Demoiselle, and I felt privileged to observe one of this species' particular behavioural quirks when a male came to land on a mat of floating long reed-like leaves very close to the bank. He was resplendent in the sunlight, blue-green body shining between blue-veined wings, the tip of his tail curled jauntily upwards. Then a female came down to join him and the pair sat side-by-side, their bodies perfectly aligned. The female curved her bronze-green abdomen slightly to one side so it dipped into the water and began to lay her eggs. While this was going on, the male stood alongside, just a centimetre or two away. Periodically he would jump up and flit around her before settling again very close by.

This behaviour is known as mate-guarding, and many Odonata practise it to some extent. Because a new male could displace the original male's sperm after mating, it is very much in the original male's interests to make sure this doesn't happen, and that no other male gets anywhere near the female until her eggs are safely laid. In most damselflies and the darter dragonflies, the male simply doesn't let go of his mate while she lays her eggs, so no other male can get the necessary headlock on the female that precedes mating. But this carries its own risks — both members of the pair are extra-vulnerable to predators when attached to each other, especially when one of the pair is half-immersed in water. Standing guard seems to be a fair compromise.

The pair of demoiselles made a wonderful picture, resting on the vivid green leaves. The process of egg-laying went on for a long time, allowing me to try a few different shots and try to get both of them in focus (nigh on impossible). The picture I would really have liked, a head-on view of their two little faces side-by-side, would only have been possible by wading into the Thames but that was a step too far for me. Nevertheless, they provided plenty of distraction, while Rob passed the time by staring out over the slowly flowing waters of the river, ostensibly watching out for clubtails but in reality sliding into a semi-coma of sleepiness.

He came back to life in a hurry when a Red Kite came flying our way, only about five metres above ground level. We had seen a fair few kites already but they had been much higher, and so much less photography-friendly. I swung my lens up and started to fire off shots, as the kite carried on towards us with lazy wingflaps. It was a breathtakingly beautiful and, at this range, huge bird, with long chestnut-and-white wings tipped with the black 'fingers' of its outer primary feathers, and that long, deeply forked tail that makes for such a dramatic and unmistakeable silhouette. Its pale eyes scanned us with interest, though not as much interest as it would have if we were lying down and looking like dead bodies — this bird is much more a carrion-eater than it is a hunter.

For the local crows, it didn't matter what the feeding habits of the Red Kite were — they saw a raptor-shaped object and they didn't like it. A couple were in hot pursuit of this kite, and as I took an overhead photo I saw the head and shoulders of a brave and angry Carrion Crow appear just above the head and shoulders of the oblivious kite. A moment later the crow must have made contact, perhaps with a kick to the kite's back, because the kite wobbled dramatically in its flight and twisted its head around to shout abuse at its aerial assailant. This had little effect on the crow, which continued to chase and harass the unfortunate kite until both birds were out of sight.

Mobbing is a behaviour that seems very puzzling and almost crazy at first glance. To go right up to a much bigger predator, even one as relatively wimpy as a Red Kite, to get in its face and generally irritate the hell out of it seems like a good way to get yourself killed. While a Red Kite would probably never pose a serious threat to a Carrion Crow, mobbing is also seen from smaller birds towards seriously dangerous predators. I've seen Swallows diving at flying Sparrowhawks, a swarm of Starlings chasing a Marsh Harrier, and Magpies having a pop at a Buzzard. In general mobbing is mostly observed in birds, but I have seen behaviour that I would interpret as mobbing in Odonata too – damselflies making little rushes at a large hawker dragon when the latter is resting or egg-laying.

There is some common sense behind mobbing behaviour. For a start, birds of the 'prey' species are quite good at determining whether a predator is in hunting mode or not. A Buzzard on the ground is unlikely to be actively hunting anything larger than earthworms. Sparrowhawks rely on the element of surprise, so if one is flying high in plain view it is probably not about to make an attack. And as far as the dragonflies go, hawkers hunt in flight, using their straight-line speed and quick turns to chase down their prey, so when they are sitting down they pose little immediate threat. But why, even if they pose little threat at that given moment, do prey species bother to take the risk of chasing and hassling their predators? They do it just because they want the predator to move on before it *does* switch to hunting mode, which will happen sooner or later. Sometimes sooner than expected – it's not unknown for a raptor that's being mobbed to turn the tables and opportunistically grab one of its tormentors.

After that moment of excitement, we returned to staring gloomily at the dragonfly-less river for a while, and then decided to take a walk upstream, just in case all the clubtails were just around the next meander. As we walked, I was briefly diverted by the sight of an unusual-looking Rabbit. I took a few photos before it bounded away, just to prove I wasn't seeing things, and the pictures confirmed it — the Rabbit was orange. Or ginger if you prefer. I've seen the occasional black Rabbit before but never a ginger one. It was small with the shortish, rounded ears of a youngster, and its fur was a striking light reddish colour, quite different to standard bunny brown, and as far as I could make out had the usual dark eyes. Perhaps it was an escaped (or deliberately released pet), or perhaps a spontaneous genetic mutant born to ordinary-looking wild Rabbits. Either way, it glowed like a little beacon and that can't have been a good thing for its long-term survival prospects. Even now, on the other side of the river a Buzzard was cruising, drifting in slow arcs on a thermal and searching the ground below for small, furry prey.

The walk upstream having proved fruitless, we wandered back to the railway bridge. The day was going downhill fast, with cloud cover increasing and sunny intervals growing ever shorter. So we spent that last half an hour on the bankside, waiting and watching. While we didn't see any clubtails we did see something else.

It was Rob who spotted her, and guided my gaze to an unopened water lily flower bud that poked out of the water like a small green mushroom. I peered at this and saw what he was talking about — a damselfly, hanging onto the stem on the river-facing side and dangling a good two centimetres of its abdomen in the water. The view wasn't great but I could see that the damsel was decked out simply but tastefully in black and cream stripes, and had a remarkably wide head and inflated-looking, very hairy white legs. She was a White-legged Damselfly, the first I'd seen for years.

This damsel, a rather localised denizen of slow rivers in southern England and east Wales, is unique among British species, the only representative of the genus *Platycnemis* to live here. The name means 'flattened leg' and describes the wide tibias (middle leg joints) of these damsels. The White-legged's species name, *pennipes*, elaborates on the same theme, meaning 'feather-footed'. I presumed this individual was laying eggs, as I couldn't think of another good reason for a damsel to be sitting with her bottom in the water, and my book said that this species often lays its eggs around the flower heads of water lilies. However, the book also said that pairs remain in tandem while the female is laying – and this one was unescorted.

Rob borrowed my big macro lens to take some photos of the damsel, and lay flat on the riverside and wriggled forwards until he was part suspended over the water. In this precarious position he took a few shots of the White-legged Damselfly, but she was unsettled and moved on, fluttering feebly off to settle on a decaying lily leaf nearby, giving us a clearer view. Here, once again she dunked her bottom into the water. Had we discovered a hitherto unobserved bit of Odonata behaviour? I had a look online later and couldn't find any reference to females ovipositing alone. Maybe she wasn't ovipositing at all. But on the other hand, what would a female do if she was busy egg-laying and the male stuck to her head was suddenly plucked away by a passing predator? Presumably she would be obliged to just carry on.

As the sky grew greyer and darker, we started to pack up our things. I was feeling pretty disconsolate despite the unexpected treat of the White-legged Damselfly. Any hopes we'd had of seeing all of Britain's Odonata in a single year were gone. There was no point making another attempt to see Common Clubtail at this site in 2011, we were already almost certainly too late and with each passing day our chances would decrease. I was still eager to meet a clubtail in person, but it would have to wait – for about 11 months.

There was a final treat though. Walking back to the car, I noticed a Great Crested Grebe swimming about in the river. Though we'd seen several over the day, this one had a passenger — a baby Great Crested Grebe — riding on its back. While Mum (or Dad) grebe was a sleek, elegant and soberly patterned bird, the chick was a hectic, humbug-striped ball of fluff with a penetratingly squeaky voice. It nestled at the front end of its parent's back, leaning cosily against her (or his) neck and yelling in her (or his) ear at top volume. A charming sight, though I did feel for the parent grebe, given the noise it was forced to endure.

CHAPTER 11

Coenagrion haſtulatum
The Northern Damselfly

꤮ ꤯

The south-east of England is probably the best part of the UK to live if you have a thing for dragonflies and damselflies. Our home in Sevenoaks put us within easy travelling distance of the most species-rich site in the UK – Thursley Common in Surrey – and only a couple of hours from key places like the New Forest, the Norfolk Broads and, for those rarities reaching us from the continent, the south Kent coast. But that's not the whole story.

By and large, insects like the heat, but in most groups a hardy few species have adapted to life in colder climates. Among UK Odonata three species only live north of the border, and a couple more have a strong northerly bias to their distribution. To see the first three and improve our chances of seeing the others, we needed to head for Scotland. The good news was that all of the target species should be on the wing at the same time, so if we planned very well and got lucky, we could find them all in just one week. The week we picked spanned the end of June and the beginning of July.

The 'big three' that we hoped to find were the Azure Hawker, Northern Emerald and Northern Damselfly. The bonus species would be the Common Hawker, White-faced Darter and Golden-ringed Dragonfly – this last species does live closer to home as well but is easier to find in the north. After spending some time cross-referencing various different distribution maps, I concluded that the best site for a Scottish base would be in Speyside, right on top of the best place to find Northern Damselfly.

I thought it likely that the only damselfly among our target species had the potential to be the most difficult of the lot, as dragonflies will continue to hunt in less than perfect weather, while damsels tend to just disappear the moment the sun goes in.

I booked a holiday cottage right on the edge of Abernethy Forest, a huge expanse of Caledonian pine forest just up the road from Aviemore. The forest, managed by the RSPB, is famous for having hosted the first nesting Ospreys in Britain for decades, back in 1954. Ospreys are now re-colonising many parts of the UK, but the pair at Loch Garten in Abernethy Forest remain probably the best known of all. The forest is also well-known among wildlife-watchers as home to a fabulous range of Highland wildlife including Crested Tits, Red Squirrels, Pine Martens, Scottish Crossbills, Capercaillies — and, less famously, Northern Damselflies. The cottage was one of a group of six known as the Dell of Abernethy, run by the superlatively helpful Polly and Ross Cameron. It had two more bedrooms than we needed, so I asked Susan if she'd like to join us for the week, which she did.

There are several ways to get to Speyside from Kent, none of them appealing. We could fly to Inverness and then hire a car. We could take a train to Inverness and, again, hire a car. Or we could drive all the way. I am no fan of flying, and an all-day train journey didn't appeal to my companions. And much as I've always longed to travel to Scotland by sleeper train, the cost was agonisingly steep. With three of us travelling together, two with a lot of camera-related luggage, it made sense to drive.

So early one Saturday morning we stuffed the Passat with our belongings and ourselves, and drove north, making a detour en route to deliver our cat to the local cattery. Rob and Susan argued over whether music should be played or not — eventually they reached a compromise whereby half an hour's music for Rob was interspersed with half an hour with the stereo off for Susan. Breaking the journey up this way seemed to make the time go more quickly, so I was all for it. As the shortest-legged person in the car, I had the back seat to myself and stared out the window as we traversed mile after mile of motorway, and began to compile a bird list for the trip. Magpie, Herring Gull, Kestrel, Pheasant, Carrion Crow … but I was already dreaming of the more exotic fare that would hopefully come our way once we were in the Highlands.

It wasn't until we'd threaded our way between Glasgow and Edinburgh that the journey started to become really enjoyable. Beyond Perth, and onto the A9 for the final stretch, the scenery became seriously beautiful, with steep rocky valleys and, to our left, the glittering Spey river running fast and shallow along its gravelly bed, drifts of angelically white Common Gulls floating over the water. Soon the great rugged mass of the Cairngorms came into view, the highest north-facing slopes still snow-covered. A right turn off the A95 took us past the pretty steam-train station at Broomhill and through dazzlingly green sheep pastures down into the little village of Nethy Bridge.

We followed a long one-way street into the shadowy edge of Abernethy Forest, and found the Dell of Abernethy cottages tucked away among tall pines. To my joy, South Dell looked out onto a fully stocked bird feeder hanging from a small birch tree, set in a sweep of lush lawn, and as we drove up to the cottage half a dozen Siskins and a resplendent juvenile Great Spotted Woodpecker with a bright scarlet crown scattered from the feeder.

The Highland evenings seem to go on forever — it was 11pm before the sky became properly dark. With maps and field guides open on the table, we planned our strategy for the next day. Susan, not so keen on dragon-hunting all week, had decided to hire a car for herself in Aviemore. Tomorrow Rob and I would take her there to collect it, and then we would go in search of Northern Damselflies.

I got up early the next morning and through the kitchen window took photos of the many birds coming to feast on free peanuts. In Kent, a scene like this would probably involve House Sparrows, Greenfinches, Blue Tits

and Great Tits, and perhaps a cheeky Grey Squirrel. Here
the most numerous birds were Siskins and Coal Tits,
and when a squirrel arrived, sending the birds fleeing
in all directions, it was not grey but red.

The squirrel bobbed up onto the low stone wall
that bordered the garden. It struck an absurdly cute upright pose for a
moment, little paws clasped as if holding an imaginary handbag under its
chin, and then it skipped along the wall, shinned up the little birch tree
and lowered itself down onto the feeder. Here it hung upside-down, all
four russet limbs wrapped around the feeder, and painstakingly extracted
peanut fragments from between the mesh, while I happily took photo after photo. The squirrel was a rich dark red, deeper-toned than a Fox,
except for its lovely long plume of a tail which was white-blonde, and its
neatly demarcated white chest and tummy. Contrary to the usual pictures
we see of Red Squirrels, it did not sport a set of pointy ear tufts — these
are a winter trait.

I was thrilled to see the squirrel, but after some 20 minutes of watching it nibbling away, still tightly hugging the slowly rotating feeder, I was
ready to move on. There was still no sign of life from the other inhabit-
ants of the cottage, but the threatened drizzle hadn't arrived, so I gathered
my stuff and set off for a walk in the woods. Though it was not yet 6am,
dawn had broken hours ago.

I've visited this woodland before a few times. It is a tremendously special place to me — I'm not much given to mystical thoughts or ideas but
can unashamedly state that to me Abernethy Forest is magical. Like many
southerners, my childhood experience of pine woodland was limited to
the sitka spruce plantations that make up some 70 per cent of Britain's
'commercial' woodlands. But while the spookily deep-shadowed interior
of a plantation does hold a certain excitement and appeal for adventurous
children, for a child more interested in wildlife the monocultured rows
of identically shaped trees, sprouting out of a lifeless forest floor with no

understorey whatsoever, offer only disappointment and boredom. Having grown up on this variety of pine 'woodland', my first walk in Abernethy was a tremendous revelation.

Here the trees are not non-native spruces but our very own Scots pines. With space to spread, each one is unique, a broad-crowned, twisty-trunked tree of character. The pure air encourages the most impressive growth of lichen I've seen anywhere, with practically every branch and twig crusted and strung with scaly, fluffy and fibrous tufts of the stuff. And the trees are well spaced, meaning sunlight reaches the forest floor and allows a thick understorey of bilberry and heather to grow waist-deep, softening the rolling contours of the ground. In the most open patches, stands of fluffy white cotton-grass cluster here and there, its presence indicating the boggy areas that serve as nurseries for some of the upland Odonata.

I took a short walk out and back along the broad trail, rutted and puddled in some places, thickly cushioned with fallen pine needles in others, enjoying the sweet air and stillness. Ahead, a Roe Deer doe stepped onto the path and paused, a burnished chestnut outline of breath-taking elegance, her great petal ears twitching uneasily as she looked my way, then turned away and trotted off. Up in the pine trees I could hear the thin calls of Coal Tits, though not yet what I was really listening for — a softly trilling Crested Tit. There would be other days though.

By the time I was back at the cottage, it was raining. Things were not looking good for a day of damselfly hunting. Rob and I took Susan to Aviemore and spent the rest of the day exploring Abernethy, while the rain came and went without a trace of sunshine to be seen. We meandered along the lovely trails to the shore of Loch Garten, and sat on rocks down there admiring the great glassy sheet of water, while Common Sandpipers whirred from shore to shore, giving their characteristic volleys of hysterical peeping calls.

At the Loch Garten Osprey Centre we went to see EJ, the female Os-

prey who has nested here since 2004. Though visitors aren't allowed anywhere near the nest, a vast pile of sticks supported on a couple of poles,
views through the telescopes in the centre are pretty good — and there was
EJ's head poking above the parapet. With her fierce stare, masked yellow
eyes, wicked hook of a bill and rather hectic tufty crest she looked half-
noble and half-comical. The contents of the nest couldn't be seen from
here, but there was a live video stream from the webcam pointing into the
nest, showing a couple of half-grown chicks hunkered down in the nest
cup. The fourth member of the family, EJ's mate Odin, wasn't around —
no doubt he was out fishing to provide for the hungry and fast-growing
youngsters.

At the end of the day, we found our way to *the* Northern Damselfly
pond, just off the roadside a little way north-west of Loch Garten. A
boardwalk led out over the small but beautifully lush little pool, which
nestled in a small clearing among pine and birch trees. Among the fine,
thready sedges were a few semi-submerged branches with photogenically
peeling bark. The setting could not have been lovelier, but it was late and
unseasonably chilly and a light rain was still falling. There were no damsels to see today, but at least we knew where to try when — or if — the sun
did shine.

The next day, things were a little brighter. We decided to spend the
morning at Tulloch Moor, an open, boggy moorland just a few miles away
that was rumoured to harbour one of our 'bonus' dragonfly species, the
White-faced Darter. This pretty little dragon is rather rare in northern
England but more widespread in the Scottish High-
lands. There used to be a tiny colony at Thursley
Common, but this was thought to have gone extinct, so the week in Scotland offered us our best
chance of finding one.

The reserve entrance was off a tiny narrow
lane that traversed wild, bleak open moorland, just

a few birches here and there punctuating the grass- and heather-cloaked undulations. While Abernethy Forest is home to the Capercaillie, largest of the UK's grouse species, Tulloch Moor is the stamping ground for its smaller but arguably showier relative, the Black Grouse, famous for its communal displays or 'leks'. Male Black Grouse gather at traditional open spots and there they do extravagant battle with song, dance and marvellously choreographed fight sequences. Driven almost crazy by a surfeit of hormones, the strutting, scrapping males are competing to impress the spectators — a handful of female grouse who assess the performances at length before selecting their favourite participant and mating with him.

Black Grouse lek early in the morning, mainly in winter and spring, and they are highly susceptible to disturbance. So our chances of witnessing lekking activity were more or less zero, but there was always a chance of spotting a lone grouse going about its business. We set off along the trail, but I stopped almost straight away when I spotted a black something flittering around low in the grass. I realised what it was straight away and called Rob over to admire one of my favourite insects.

The moth, one of several around, settled halfway up a tall grass stem. It was a small moth, which rejoiced in the name 'Chimney Sweeper'. From any distance it looked plain black but when I crouched down for a closer look I could see the little white tips to its forewings. The wings also reflected a very subtle iridescence: hidden in that blackness were hints of blue, gold, violet and green. It rested with wings fully spread, like a butterfly, until I poked my lens a bit too close and it took off again.

Nearby was another distinctive insect, this one preferring to get around on foot. A large and sturdy green beetle, it marched purposefully along the trail, and when I opened my hand in its path it climbed onto my palm without hesitation. I lifted my hand up so Rob could have a good look. It was a Green Tiger Beetle, formidable predator of heathery habitats. From head to tail it was a smart shade of deep green, and the long shield-shape of its wing-cases were marked with creamy, black-circled spots. As befitted its terrestrial way of life it was long-legged, equipped for chasing down less speedy prey, and its broad, wide-eyed head bore a pair of long, restless black antennae. Realising it was no longer on familiar ground but instead standing on something warm and squashy, it stopped to appraise the situation and Rob managed a few photos. Then I gently put it back down, and it prowled off into the heather — a tiny tiger on the hunt.

I felt encouraged by these two insect encounters. The third was even better, because it was a dragonfly — our first Scottish Odonata. It was a big one — certainly not a White-faced Darter but some kind of hawker. It swished past us on a skitter of brittle silvery wings and pitched down on top of a bulging heather clump, luckily at the edge of the trail. We approached, slowly and carefully, and were soon enjoying close views of our first Common Hawker.

The hawker, a male, was surprisingly difficult to make out against his heather backdrop, despite having a pretty arresting black and blue colour scheme. The paired blue spots that ran down his black abdomen served to break up his outline and provide what the textbooks call 'disruptive camouflage'. Once I had my eye in properly I could appreciate his finer details — the dramatically 'pinched-in' waist, the delicate yellow edging to the wings, the wonderful smoky-blue eyes. Common Hawkers are found south of the border, but don't reach down to south-east England, and this was the first one I'd ever seen.

The Common Hawker is a member of the genus *Aeshna*, which contains several large and impressive dragons now known as hawkers. Up until the 1980s the genus name doubled as the English name for most dragonflies, so the hawkers were the 'aeshnas', and this species was the Common Aeshna. But these days more utilitarian and user-friendly words like 'hawker' 'chaser' and 'darter' are used for dragonfly English names, which is a good thing on balance. The word *Aeshna* sounds quite pretty until you know what it means in Greek — 'ugly', or 'mis-shapen'. Hawker is much better. The Common Hawker's species name, *juncea*, means 'rushes', a group of plants that are generally present in profusion in good Common Hawker habitat.

The field guides say that the Common Hawker is of a nervous disposition, intolerant of close approach, and this proved to be so — we inched too close and he was off, zooming out of sight in moments. We carried on along the trail, checking around us on all sides for further dragonfly activity, but there was none. Instead, a post-breeding flock of Mistle Thrushes came over, giving their loud rattling calls, a sound more reminiscent of a suburban garden than this remote and rugged wilderness.

When the trail took a downward slope towards an expanse of obviously wet and boggy ground, I had high hopes that here we would find dragonflies, but again we drew a blank. Risking life and limb I edged cautiously out over the bog as far as I dared, but when the ground underfoot began to feel more spongy than springy I backed off, defeated, and scanned far out across the bog with my binoculars. Not a dragon in sight, and — worse — the clouds were starting to close in. We debated tactics, carried on along the trail a little further to be sure we had checked the full sweep of the bog, and then reluctantly agreed to turn back.

The walk back towards the car was uneventful until the last moment. Clouds had consumed the sky at alarming speed, and there was a touch of rain in the air. As we passed a tall birch some instinct compelled me to look up, and there against the grey sky was the outline of a small dragonfly, several metres up. We both peered up at this little flying shape rather helplessly. Then, as if swatted by some divine hand, the little dragon tumbled downwards and crash-landed upside-down on the heather right beside us.

We were too surprised for a moment to do anything, and a moment was all that we had. The dragon seemed to be stuck on its back, its wings caught up and its six stiff legs paddling away at nothing. There was no time to do anything sensible before it pulled itself free and flew rapidly away. But stuck in my mind's eye was an impression of its upside-down face — the dark edges of its eyes on either side of a frons that was pure white.

Rob turned to me. 'So. What was that then?' he asked, showing touching but sadly misplaced confidence in my identification skills. I said I wasn't sure, but the white frons and small, slight build suggested it might be the very species we'd hoped to see, the White-faced Darter. But with no photo and no view of the dragon's top side there was no way I could be sure. Already the picture in my mind was fading, distorting. Was its face really white, or just pale? Was its body slim enough? Maybe before I'd started trying to take photos of absolutely everything I saw, I'd have had a bit more faith in my tentative identification. But there was no way around it — the dragon had gone, and as the rain started to fall in earnest the chances of re-finding it, or tracking down another, had gone.

The White-faced Darter is rather different to the other British darters. It belongs to a different genus to them for a start — they are *Sympetrum* and it is *Leucorrhinia* (meaning, rather charmingly, 'white-nose' in Greek). The species name *dubia* is related to our word 'dubious' but exactly what's dodgy about this dragon is not clear. Males are deep, dark red with a good helping of black, females mostly black with patches of dull yellow. The male in particular has a slim-waisted body which, coupled with his small size, makes him somewhat reminiscent of some kind of large wood-wasp. Both sexes have the characteristic white 'nose'.

It rained for most of the rest of the afternoon and we went back to the cottage and passed the time taking a very un-Scottish siesta. By 6pm things had cleared up, and we went back to the little damselfly pool for another look — it seemed to me that it would be warm enough to coax a Northern Damselfly or two out of hiding.

In the glow of evening sunshine, the pool was more idyllic than ever. There was no-one else around — the peace was perfect. Sadly, there were also no damselflies around. It was difficult not to be discouraged — two days in Scotland were already gone and we had seen only two Odonata (well, one and a half, really). The following day, Tuesday, was forecast to be a good-weather day and we had earmarked it for a longer trip, but I decided that there would be time for another quick check on the pool en route.

On the third morning of our Scottish sojourn, I again woke hours before either Rob or Susan, and wandered off into the woods alone. I was steadily amassing a reasonable collection of bird and mammal photos, being particularly pleased with my shots of a gawky little Tree Pipit fledgling, a beautifully burnished Roe buck that I managed to get pretty close to thanks to my stealth and the fact that the buck was distracted by a nearby doe, and a confiding Willow Warbler. Today, I was deep into the forest when I finally heard that purring trill I'd been listening for from day one, and after some patient scanning through the branches found the bird responsible, a delightful Crested Tit.

On mainland Europe, 'Cresties' are quite common in all kinds of habitats. Here in the UK, they only live in ancient pinewoods in the Highlands. This seems a desperately unfair situation, but it does mean that British birdwatchers really do appreciate the charm and beauty of this little bird when they see it, as they almost certainly made a special trip for the privilege. The Crestie I found was picking about among the wafting lichen strands dangling from a snapped-short pine branch. It dangled upside-down, its spiky little crest pointing downwards, and kept up a constant stream of trills, talking to another Crestie foraging nearby but out of sight.

After breakfast, we made our third visit to the damselfly pool. We had sunshine, warmth and reasonably still air, and we were weighed down by a painfully intense burden of hope and anxiety as we walked out onto the boardwalk. Would today be the day? Yes, it would. Almost as soon as we began to look we simultaneously clocked a spindly little blue and black damsel resting on a sedge stem, and a quick look at it through my

camera revealed the presence of the '*Coenagrion*' spur on its thorax side. All *Coenagrion* damsels have this little black marking, just under the main thorax stripe, but up here in the Highlands there are no other *Coenagrions* to confuse the picture, only the Common Blue Damselfly, which is not a *Coenagrion* but an *Enallagma* — with no spur.

Closer scrutiny of the waterside vegetation revealed a few more Northern Damselflies, one of them actually resting on the pretty peeling log that I'd noticed on our first visit here. The male Northerns looked smaller and a paler shade of blue than either Azures or Common Blues, but their black markings were more extensive, especially at the tail end. The black marking on the second abdominal segment resembled a spade from a card deck, with a clear pointed tip, distinctly different to the round-topped 'ball-on-stick' marking the Common Blue carries on this segment. In fact whoever gave the species its scientific name saw the marking as a tiny weapon — the species name *hastulatum* means 'little spear' in Latin. The females wore a more generic black-and-green colour scheme.

This damselfly's distribution in the UK is extremely restricted. The field guide shows just three small green blobs, all in the north-central Highlands, with the one on top of Speyside by far the largest. A much more detailed distribution map, pinpointing all known sites, would show only two or three dozen blobs, all small pools much like this one, and each representing a colony of about 100 insects. That makes this species one of our scarcest, and a high conservation priority, though unlike its Southern counterpart, the Northern Damselfly is not protected by law. Maybe if it were, Tesco would not have been able to obtain planning permission to fill in one of those pools in 2012, in order to build a new superstore in Aviemore. Although the developers plan to create a new pool for the damsels, and to physically move the nymphs to this new pool, there can be no way of knowing the insects would thrive at the new site — similar translocation schemes involving other rare insect species have certainly failed in the past. Even in the heart of the Cairngorms National Park, the struggle between wildlife and development is ongoing.

These Northerns, under no immediate threat of eviction, were not doing very much, just sitting around resting, getting warm, and making the occasional short jaunt to a slightly more sunlit perch. Also soaking up some early rays were a couple of Common Lizards. One of these was a tiny individual, as dainty and perfect as any handcrafted dragon ornament from a hippy gift shop, looking extremely comfortable curled up on a plush bed of bright yellow sphagnum moss. Another splash of colour was provided by a Small Pearl-bordered Fritillary, a gorgeous little butterfly with a chequered pattern of black on its fiery orange wings. The fritillary detained me for several minutes as I struggled to take a clear photo of it, but to no avail as it threaded its way through the tangle of plant leaves and stems alongside the pool, finding impossibly small flowers from which to sip nectar. I had to give up in the end, and we returned to the car to head for Coire Loch in the Glen Affric area, in search of Northern Emeralds.

CHAPTER 12

Somatochlora arctica
The Northern Emerald

ℜℭ ℑℜ

Having made the acquaintance of the Northern Damselfly, Rob and I continued westwards, towards Glen Affric. This drive took us half-way along the western shore of Loch Ness, and provoked a lively discussion on the existence or non-existence of the loch's eponymous monster.

I feel a little torn on this subject. I've been along this road before, a few times now, and each time I am struck anew by the sheer, honest-to-goodness, hugeness of the loch. As you drive along the scarily narrow A-road, cut into the steep hillside, you pass mile after mile of broad and implacably deep water, and you start to suppose that there is enough underwater space there to completely hide an entire sunken city, never mind one solitary monster. Nessie is usually described as a plesiosaur – a four-finned, swan-necked swimming reptile – and the largest plesiosaur fossils ever found barely top 20 metres. If you got hold of a living plesiosaur tomorrow, and decanted her into Loch Ness, I have every confidence that she could live and thrive very happily there and quite possibly never be seen again.

The problem is that of course there can't be just one Nessie. Population biology and everything we know about animal lifespans stops that idea before it starts. Large reptiles can be very long-lived indeed – there are several tortoises on record that reached 150 years old or more. But records of Nessie go back to the seventh century. We have to be talking about not one but a whole lineage of plesiosaurs, surviving and breeding in the loch not just since the seventh century – but since the most recent of our fossil plesiosaurs died – 65 million years ago. And there would need to be enough of them to avoid genetic catastrophe from too much inbreeding. Suddenly Loch Ness doesn't seem nearly big enough.

According to Rob, today's technology is advanced enough that we would almost certainly have located one or more Nessies by now using sonar or other scanning techniques. But maybe this will never happen, just because it's nice to keep the legend alive. Conclusive proof of Nessie's non-existence would certainly be a blow for the various Nessie-themed cafes and hotels scattered along the loch side.

We steered away from the loch halfway down, turning west at Drum-nadrochit. The almost traffic-free A831 took us through delectable low-lying countryside of pasture, copse and woodland, all of it intensely lush and green after a spring of more or less non-stop rain. Approaching Cannick village, we drove over the lively river Affric, racing along on its bed of tumbled stones. Soon after that, it was off down a single-track road which took us alongside the river, then alongside the lovely Loch Beinn a' Mheadhoin. The scenery now had become more rugged, with steep rocky slopes rising on the non-loch side of us and ferns and spindly little silver birches sprouting out of the gaps between the lichen-plastered granite outcrops. We were deep into Glen Affric now, apparently the most beautiful of all Scotland's glens, and the road began to climb up away from the shining loch. Here we found the starting point for the route that would give us access to the footpath to Coire Loch, home of the Northern Emerald dragonfly.

Rob hadn't even shut off the car's ignition when there was an officious little tap on the car bonnet. We both looked up in surprise, just in time to see the tapper – a male Chaffinch – flit from bonnet onto wing mirror. Here he sat expectantly, bright little eyes scrutinising the interior of the car for anything that looked like food. The only thing I'd brought to eat was a cereal bar, so I crumbled a bit of that into my palm, got out and offered my outstretched hand to the Chaffinch, which flew over immediately to alight on my fingers and start tucking in. Soon a few other

Chaffinches had come over to see what was going on, and were eyeing up the cereal crumbs.

Where I come from, Chaffinches aren't this friendly. Quite the reverse – they are jumpy and furtive, and you stand a much better chance of coaxing a Robin or a Great Tit to come and eat from your hand. Their rather timid ways make them far less familiar to the average human than other, less common garden birds, despite the male's showy blue, pink and pied plumage. I remembered years ago seeing a male Chaffinch feeding outside an aviary full of colourful tropical birds at a zoo, and all the on-lookers were convinced it was an exotic escapee, on the wrong side of the bars. But I was also reminded of the car park at Ladybower Reservoir in the Peak District, where visitors are besieged by large flocks of super-confident Chaffinches. Make the mistake of buying a pasty from the little tea-shop and if you don't retreat into your car to eat it you'll be thronged with hungry and fearless little finches. The Chaffinch is obviously a versatile and enterprising avian – not so surprising, really, when you consider it is Britain's second most numerous bird species.

I didn't let the Chaffinches eat the whole cereal bar – we had a long walk ahead of us. A signpost at the car park edge indicated a number of colour-coded trails. One, indicating the yellow trail, pointed towards Coire Loch. Following it, we set off alongside the river. Up here, it was a fierce little torrent, roaring along at the bottom of a deep gorge with steep banks of bare rock. The gorge became shallower as we went along, and the river a little wider and slower. Rocks of various sizes interrupted and diverted the flow. The whole set-up looked very good for Dippers – charismatic, dumpy little birds that specialise in hunting water creatures in fast-flowing, rocky rivers. I love Dippers and very much hoped to see a few on this Scotland trip – they don't occur down in Kent – but the river proved a Dipper-free zone, today at least.

The trail veered away from the river after a mile or so, and took us across a tiny road and up into a patch of birch woodland on a precipitously steep hillside. We'd left the crowds behind as well as the river. As the path carried on up, and up, and up, I started to think that maybe the casual crowds had the right idea. We emerged from the woodland onto open moor, all tough knee-high heather and bracken. The trail wound upwards a little further, then levelled off and we stopped, a little breathless, to look out at a truly glorious view.

The hillside fell away and the splendour of the glen was rolled out before us — heathland and pine forest on the rippling slopes of the deep valley, flanked by brooding blue mountains in the distance. The sky, patchy with cloud, cast a pattern of sun and shadow across the sweep of countryside. We couldn't see a single road or building — for miles in all directions there was scarcely a sign of human life. And at the lowest point of the valley, centre stage in all of this was Coire Loch itself, a small, nearly circular loch of deep, shining blue, its sphagnum-mossed banks brilliant yellow, looking like a sapphire set in gold.

This loch is — I think — a landscape feature called a 'kettle' hole, the result of glacial activity. As a glacier retreats, sometimes lumps of ice fall from it and then get buried by fast moving flood water, which forces them into the ground. As things warm up further, the ice lump melts and leaves a flooded hole in the landscape. Being shallow with a wide surface area, Coire Loch gets lots of light and warmth, encouraging water plants to grow. It is, therefore, very different to the kind of loch that forms when a glacier carves out a deep U-shaped valley. Loch Ness is a good example of the latter — it's long, narrow and extremely deep and cold. So Coire Loch is a vegetation-rich little haven for dragonflies, while the much larger Loch Ness is no good at all to Odonata-kind.

We stayed up there a little while, taking photos. Each was different, as the configuration of cloud was constantly changing — in some the loch shone sky blue, in others it was shadowed and looked like a bottomless pool of ink. Finally, when the clouds were becoming more numerous and it looked like we might run out of sun altogether, we carried on down the trail, which took us down the steep hillside towards the loch. As we went, we disturbed a big dragonfly that had been resting unseen on the heather — but it shot away before we could gain any impression of it beyond 'big' and 'very fast'.

As we got closer, we noticed two interesting things. The first was a plump little water bird, sitting on the loch. My thoughts went wild pondering what it could be — surely something uncommon, in such a wild place. The other was easier to identify — a man standing down there near the loch shore. Could this be another dragonfly-seeker? The Canon camera and massive lens he was carrying suggested that maybe he was.

Down at the shore, I got my binoculars on the little water bird and saw to my disappointment that it was a Little Grebe — common as you like. It was not alone either, lurking among the shoreline sedges were its mate and a couple of well-grown chicks. The grebes made an incongruous sight on this little loch, miles away from other water. I don't think I have ever seen a Little Grebe in flight — they swim and dive with aplomb but with their stubby wings and rotund bodies are ill-equipped for aerial activity. But these two adults must have flown here at some point. Or maybe their parents did, or their parents' parents ... but the progeny of the incumbent pair would certainly have to fly away from the loch. There wasn't enough room for more than one couple.

The man by the loch shore gave us a nod of greeting as we wandered his way. Not wanting to intrude, we stood well back and soon realised that there were dragonflies here, lots of them, flying along the edge of the water and chasing each other. I could see that they were all emeralds, but that was as far as I got. According to the book, all three British species occur here — Downy, Brilliant and the one we particularly hoped to see, the Northern Emerald. Brilliant Emeralds are the easiest to identify, the males a particularly dazzling iridescent green that catches the light even when they're some way off. But the other two species are both pretty similar. None of the individuals I could see were in any way brilliant, so I figured we were looking at Downies, or Northerns, or both.

Having convinced myself that the man with the Canon was an expert in all things emerald, I squelched across the increasingly waterlogged sphagnum to join him and asked 'Do you know if these are Downy or Northern Emeralds?' He promptly shattered my illusions by laughing sardonically and replying 'No idea', but then produced his field guide — the same one that I had — and together we pored over the diagnostic features of the two species. Male Northerns have claspers that are shaped like callipers, while those of the male Downy are straight. Which is all well and good, but you try detecting that when the dragonfly in question is flying past at some distance and high speed. The other key feature, the facial pattern, is similarly difficult to observe. Canon man and I agreed that the only chance of making a confident identification was to find an emerald in repose.

In fact the two most visually similar species, the Downy and the Northern, are actually classed as belonging to different genera. So for all their apparent shared features, they're not that closely related. The Downy's genus, *Cordulia*, contains just three species, the other two being American and east Asian. The Northern is one of a few dozen species in *Somatochlora* — so is the Brilliant Emerald. There are *Somatochlora* species in boggy and swampy countryside across Europe, Asia and North America, including the Clamp-tipped Emerald, the Ski-tipped Emerald and the Brush-tipped Emerald. It would seem from these names that 'tip shape' (which I'm guessing means the claspers) is rather crucial in separating the various species. The actual genus name means, unexcitingly, 'green body' in Greek, while the Northern Emerald's species name *arctica* reflects its northerly distribution.

I rejoined Rob and gave him the rundown on the discussion. Tasked with tracking down a dragon on the deck, he went off to the southern side of the shore, while I stayed put, watching the emerald dragons dashing around. In typical emerald style the males were patrolling their territories, making short dashing flights before hovering, and whenever another emerald came near they would give chase. Among them, a female or two were flying low over the water, dipping in their abdomen tips to lay eggs. To me, they all looked pretty much the same as the Downy Emeralds that I'd seen for the first time ever on my local patch just a few weeks ago, but it's hard to tell from a field guide just how different two species will look when you encounter them in the flesh. Sometimes, two that look almost identical on paper are easy to tell apart in the field, because of subtle little differences in the way they move that add up to a distinctly different impression. Maybe when I did see a Northern I'd immediately recognise it as 'something different'. But then again, maybe I wouldn't.

There were other Odonata here besides the emeralds. In fact, the species list for the loch is 14 — very impressive considering how small it is, and how remote it is from any other suitable habitat. Of those 14, a few were in evidence today. I spotted Large Red Damselflies, Four-spotted Chasers and a few Common Blue Damselflies. The last one of these was particularly interesting. Compared to the male Common Blues from down south, the males up here had much more extensive black markings on their abdomens. The blue bits looked different too — a deeper and brighter shade. Checking the book later, I could find no mention of this regional variation, though I did learn that Common Blues actually turn grey when they are very cold, and that in the Highlands it can take up to four years for a Common Blue larva to become mature, compared to just one for the same species down south. Maybe the prolonged maturation has some relevance to the colour of the adult insect, or maybe extra black coloration helps them to warm up more quickly when the sun comes out. There was no doubt about the identification, in any case.

Another intriguing sight was a submerged water lily flower, which had attracted a huge assemblage of tadpoles. They pressed themselves against the decaying cream-coloured petals, squeezed between them, and generally nestled around the flower as if it was their comfort blanket. Most of the tadpoles were quite well grown with little spindly hind legs kicking away either side of their shrinking tails, but I couldn't be sure if they were destined to grow into Common Frogs or Common Toads. But suddenly the lily lost its appeal and the little swimmers all dispersed.

A shout from Rob now brought me over to inspect a male emerald dragon which had stopped whizzing about and was taking a breather on the carpet of sphagnum. At last, the chance for a proper look. The dragon was evidently tired out from a hard morning's territorial defence and let us get pretty close – close enough to see that his claspers were, sadly, not calliper-shaped. He was a Downy. Still, he was the first Downy I've ever seen perched, so it was good to get a close look.

The emeralds as a group look, to me, more primitive than any other dragonflies, and this individual was no exception. Each abdomen segment seems to slightly overlap the last, creating a rough, ridged look. The head, thorax and abdomen look loosely connected, as if the whole creature could fall apart at any moment. Even its wings, with their heavy and obvious dark venation, looked like cut-price versions compared to the finer, more elegant-looking wings of other dragons. In my mind's eye, the gigantic pseudo-dragonflies that flew over the Carboniferous landscapes of earth a few hundred million years ago were monstrous versions of Downy Emeralds. The little dragon's best features beyond doubt were its fabulous bright green eyes – the rest of its body was a much more subdued blackish-green, covered with a soft fuzz of white-blonde down.

Rob wandered away a short distance, and found a pair of emeralds joined in the copulatory wheel position, clinging to a frond of heather. With no regard for their privacy I too went over for a closer look. It wasn't possible to inspect the male's anal appendages, as they were being deployed in their intended purpose, maintaining a vice-like grip on the female's neck, which looked like it was about to snap under the strain. However, the male's little green-eyed face was upturned towards me, providing a not-to-be-missed opportunity to check the pattern against that shown in the field guide.

The field guide helpfully includes head-on illustrations of the three emerald species' faces. Downy has no yellow on its frons, Brilliant has a lot, and Northern falls in between, with just two yellow spots on the sides. I peered at the male emerald and there they were, two distinct spots of sparkly, speckled yellow, like blobs of gold glitter. We'd found a Northern — in fact two Northerns (assuming that the female was the same species as her partner). Moreover, the couple seemed set for a long interlude of love making, so we could examine them at leisure.

Emeralds spend about an hour 'in cop', during which they appear to be sitting still and quiet. The female, her neck bent forward in an un-comfortable-looking position, was hanging onto the male's abdomen with all six feet — the bases of the male's hindwings are 'cut away' slightly to make this possible. Her own abdomen was curved forwards so the tip of her body could make contact with the underside of the male's second ab-dominal segment. What was going on at the point where their two bodies met could not be seen, but I have it on good authority that the male would be using his dragonfly equivalent of a penis to push any previous male's sperm out of the way before depositing his own on the female's eggs. Who says romance is dead?

The walk back was a big anticlimax after all the action at Coire Loch. We didn't want to leave, but the sky was darkening and the dragonflies were making themselves scarce. As we climbed the return loop of the trail, it began to rain. The only creature of interest that we saw was a Common Blue Butterfly, sitting uncomfortably on a spiny thistle head with its wings tight shut, conserving warmth. Back at the riverside, I found a boulder to sit on and immersed my feet in the breathtakingly cold water — the walk hadn't been a particularly long one but the gradients were tough and my toes were sore. It was mid-afternoon, but too late in the day for further dragon hunting, so we made our way back to the cottage and spent the evening sorting through the day's very satisfactory haul of photos.

CHAPTER 13

Aeshna caerulea The Azure Hawker

℞ ℈ ℞

Aﬁter the successful dragon hunt at Coire Loch, we had just one more planned trip and three days to do it. With poor weather forecast for the western Highlands, on the first day we headed for the coastline east of Inverness to look for other wildlife.

After arriving at the small town of Banff, we made for the harbour and sat on the wall to relax. Sunshine and almost warm air had a spectacular effect on our spirits, as did the constant pageant of seabirds, many of them flying past right up close. There were Kittiwakes, Eiders and Guillemots, and further out to sea Gannets were feeding. This was enthralling to watch. The Gannet circles high over a likely spot, then tilts downwards, starts to drop, and as its dive nears the water it draws back its wings to turn itself into a feathered harpoon that punctures the dark waves with a neat, round splash.

Later we moved onto Troup Head, an RSPB reserve. The cliffs here are home to a huge seabird colony — all those Gannets, Kittiwakes and Guillemots at Banff would probably have come from here. Idyllic quiet lanes through peaceful, gently rolling farmland led us to the tiny car park, and then a short trail through wild grassy heathland to the cliff top, and the sensory wallop of a seabird colony in full swing. The intensity of noise and smell was extreme, and as we reached the cliff edge and found a place to sit, it became clear that the cameras' memory cards were in for a busy day. Fulmars, Kittiwakes and Gannets went constantly to and fro at eye-level. The long, narrow ledges were packed with Guillemots, with Razorbills on the wider bits, and on the lowest rocks Shags sat elegantly upright, their black plumage bearing an oily green lustre and their heads comically adorned with backcombed quiffs. A few Puffins and the occasional Great Skua completed the picture. The sun lasted well into the afternoon. On the walk back we saw dozens of portly, black-and-scarlet burnet moths on the heath, most of them coupled up, and a single Dark Green Fritillary butterfly.

It had been a wonderful day, but there was just a Thursday and a Friday to go before home time and one important dragonfly search still to do. Neither day was looking good weather-wise. But time was running out, so we hung our hopes on the BBC's promise of a few sunny intervals in the afternoon on Thursday, and that morning we began a westwards journey across the slim waist of northern Scotland.

The drive traverses some beautiful and spectacular scenery. On the far side of Inverness, we looked out for Red Kites as we passed near the Black Isle, and saw a couple cruising by the roadside, all long three-toned wings and twisty forked tails, looking out for some squashed delicacy. Here the landscape is pastoral and very suitable for Red Kites, which were reintroduced here in 1989. Sadly, these kites have not thrived to anything like the extent they could have. In 1989, reintroduction schemes were began here and also in the Chilterns, and identical numbers of young kites were released. By 2006, there were 320 breeding pairs of kites in the Chilterns, but only 49 pairs on the Black Isle. The problem seems to lie with a minority of landowners who think that the kites might pose a threat to livestock or game (wrongly — kites mainly eat carrion), and so shoot, trap or poison them. Deliberate killing of this kind is still the biggest problem for birds of prey in Britain — decades after it became illegal.

The A832 describes a large and almost complete loop around a picturesque chunk of the central and western Highlands. We had explored the full length of it a few years before, on a holiday based near Gairloch on the west coast. It felt good to be back, among rugged hills and mountains and the occasional still loch or lively river, and better to be heading for the lovely Loch Maree. The road is lined with stands of fiery yellow flowering gorse as you approach the stunning large loch along a deep valley, tucked in between the wild, forbidding heights of the Beinn Eighe mountains. Only the sky, which was very unsettled, gave cause for concern.

On the roadside along the shores of Loch Maree, there is not a gift shop or a fibreglass monster in sight. However, the Loch's own legendary monster, the snake-like 'Muc-sheilch', is surely as real or unreal as Nessie. Deep enough to conceal dozens of cryptozoological phenomena, Loch Maree fills a long, steep valley in the rollercoaster landscape of the western Highlands, beginning under the shadows of the Beinn Eighe mountains and emptying into the Atlantic 12 miles later at the seaside village of Poolewe. At the loch's widest point, a jumble of pine-covered islands offer safe nesting places for Black-throated Divers, which are among Britain's rarest breeding birds.

We followed the loch shore for several miles, and stopped just beyond the Bridge of Grudie, where the shallow and stony River Grudie feeds into the loch from the south, parking with some difficulty on a steep and narrow ridge. A plateau of heather apparently stretched away to the hills in the south-west, but the waving crowds of fluffy white cotton-grass heads revealed that this carpet of dark green was really just a collection of islands and bridges forming uncertain pathways across a sheet of black, boggy water. We crossed the road, climbed over a stile and then set off onto the bog, picking gingerly from tussock to tussock and feeling the ground compress disconcertingly, like soft rubber, under foot. Overhead rolled a non-stop parade of clouds as huge, complex and rugged as the hills themselves.

For a place so wild, there seemed a remarkable lack of wildlife. The occasional shrill peeping of a passing Meadow Pipit broke the quiet now and then, but the foreboding skies were empty of birds. Stopping to peer into one inky pool I found a Palmate Newt, delicate and almost translucent. This newt, the smallest British species, is the only one tough enough to do well in the far north of Scotland. Looking around at the scenery, grim

and chilly even on this July day, I couldn't even imagine what it is like in January — and a newt can't even move unless it's reasonably warm — it would be a tough life indeed.

Tough too for the insects. Nymphs of the big dragonflies here take years to reach maturity, spending months and months completely inactive when the weather is coldest. Nevertheless, the three species that occur here do so in good numbers. They are the Common Hawker, Golden-ringed Dragonfly, and the one we hoped to see most of all — the stunning Azure Hawker, one of Britain's scarcest dragons and, because of its liking for remote and somewhat hazardous habitats, probably the species least likely to be 'bumped into' by the more casual wildlife-watcher.

Male Azure Hawkers are like super-blue versions of Common Hawkers, though are a bit smaller and stockier. It will be no surprise when I reveal that their species name *caerulea* comes from a Latin word meaning … bright blue. While the Common Hawker looks black, and reveals its blue spots when you get a clear look, the Azure has more blue than black on its body and so looks blue with black spots rather than the other way around. According to my book, the Azure is not shy like the Common Hawker, but is curious about people and will approach closely to 'buzz' them. What an experience that would be — to be dived at by one of the rarest dragons in Britain.

The clouds kept rolling in from the south-west, until at last a break appeared, sending a band of sunshine spreading across the hillside, heading our way. It came slowly, taking its time to negotiate the tricky terrain on the way. In its wake, more cloud. When the break reached us, it would only last a few minutes. But it took moments, not minutes, for the sun's warmth to work its magic.

Soon after the sunshine hit our faces, a dragonfly rose up between us from its hiding place and with a clatter of bone-dry wings looped around us. I had an impression of flaring, vivid blue light, like a strange-shaped

Kingfisher, as it passed, and then it was racing away south towards the mountains, putting metres and miles between us in seconds. Neither of us had so much as lifted our cameras. I called to Rob, who'd had a closer and longer look than me, 'Did you see that? It looked very blue?' 'It was *made* of blue,' he called back.

This was, I felt sure, an Azure Hawker, in spite of its behaviour being exactly the opposite of what the book said it would be. There was no time to dwell on the disappeared dragon, though. As the sun warmed the ground, more and more dragonflies appeared from here and there in the heather. A dark, slim, biggish dragon landed on a lichen-covered rock and we crept over to take its photo. It was a Common Hawker, a male. Small blue spots picked out each segment on its long abdomen, which was parallel-edged apart from a wasp waist just behind its powerhouse thorax. Over by one of the larger bog-pools a very big dragon slammed its brakes on in mid-air and hung there, stabbing fiercely and repeatedly at the water with a curved-down abdomen. To my delight it was a Golden-ringed female, the third of the 'big three' here. This is the longest-bodied of all Britain's dragonflies, thanks to the female's very long spear of an ovipositor. She was laying a batch of eggs in the cold and murky pool. The resultant nymphs would take up to five years to mature, growing infinitesimally slowly in the black, chilly water. We tracked her down when she flicked away and settled nearby — a green-eyed beauty, her graceful black body accessorised with a chain of flashy gold bands. Nearby, a trio of Four-spotted Chasers lived up to their name, dashing irritably at each other in an argument over territory.

After this, Rob ventured off across the bog on his own to take some photos of the River Grudie. The river would provide the necessary pretty foreground for photos of this brooding landscape, under the complex kaleidoscope of cloud. I watched him go, feeling mildly anxious as he took huge steps from tussock to tussock, tripod swinging about as an aid to balance. Then over a big bump and he was out of sight. I carried on looking for more Odonata. The sun came and went behind advancing banks of cloud, but there looked to be a good long interval of mainly sunshine ahead.

These boggy uplands must be all over the place in the Highlands, but most would be difficult or impossible to access safely. A dryish path of sorts traverses this particular bog, but to get near the larger and more interesting pools it was necessary to venture out onto the soggier ground. It was worth it though, for each pool was a hive of activity. The Four-spotted Chasers were the most numerous species by far. These tough mid-sized dragons are one of Britain's most widespread species, and in good habitats can be present at high population densities. Every pool out here held several, either perched on favourite rocks or sticks, or energetically attacking each other. There were also a few Common Blue Damselflies which were, like the ones we'd seen at Coire Loch, much darker than those back home.

I found another female Golden-ringed, this one perched and fairly tolerant of my close approach. She was a most spectacular creature. Looking at her from the front, I could see that her glorious bright green eyes were different to those of a hawker — rather smaller and wider, and they only met at a point near the top, rather than showing the full wrap-around effect. The long ovipositor looked in good condition, though the field guide informed me that over days and weeks of use it is liable to become damaged. The rather violent stabbing motion used during egg-laying, which helps embed the eggs in stream beds so they aren't washed away, eventually takes its toll on the poor female's delicate anatomy — the lower abdomen also tends to become discoloured after prolonged dunkings in muddy water. This individual was pristine though, and looked wonderful, her vivid black and yellow colour scheme giving the impression of a huge, elongated wasp.

The Golden-ringed Dragonfly is the only British species in the genus *Cordulegaster* (though there are many others, all rather similar-looking, in Europe) and indeed in the family Cordulegastridae. We have come across a similar scientific name before – *Cordulia*, the genus of the Downy Emerald, and the words are both derived from the same root, the Greek word for club-shaped. The 'gaster' part comes from the Greek 'gaster', for stomach. So this lovely dragonfly and its relatives are named for their 'club-shaped stomachs' – not the most flattering ephithet. Nor is it very apt for the females, whose bodies are very straight up-and-down, it is the males who have the club-shape. The species name *boltonii* honours an 18th century naturalist called Thomas Bolton, who collected the first specimen.

Mr Bolton must have had good dragon-finding and -catching skills, better than mine, at least. In my efforts to get good photos of this Golden-ringed Dragonfly, I must have moved too suddenly or got too close, because the lovely dragon was suddenly off across the bog, and there was no way I could safely chase after her. I started to wander back to the start point, and, turning around, saw that Rob was on his way back too. He followed what must have seemed to him to be the most direct path across the raised bumps and ridges, but from where I was standing looked amazingly circuitous and painstaking. As he got closer, I could see that his trousers were thoroughly soaked up to mid-thigh height.

'I fell in the bog,' he called ruefully, when he was within shouting distance. 'Is the camera OK?' I called back (well, it is much easier to dry out a human than a camera). He said it was fine, he'd held it clear when he'd mis-stepped into a rather deep pool. I said he was lucky it hadn't been deeper. But he looked unhappy, smelled extremely boggy, and it was clear that we'd not be hanging around here any longer.

The series of sunny intervals was coming to an end in any case. The next wave of cloud was fast approaching and looked very solid. As soon as the lights went out and deep cloud-shadow engulfed the landscape, every dragonfly vanished. We waited a short while, to help Rob dry out a bit before getting back in the car, in a forlorn last hope of seeing another Azure Hawker, but we didn't see any more dragonflies, let alone the astounding bright blue creature. We had to leave without a photo of this most special dragon. Instead we brought back photos of its extended family, a pair of trousers in need of a good wash and a trio of engorged deer ticks attached to my lower legs that would cause an evening of reading up about Lyme disease followed by weeks of vague anxiety. But the memory of our one brief Azure sighting stayed with us, a bright blue slap across the retinas against a landscape of grey cloud, grey mountains, grey-green moorland and pitch-dark water.

CHAPTER 14

Sympetrum danae The Black Darter

❧ ⁂ ☙

I was keen to get to Thursley Common, England's premier Odonata site, at least once in 2011 to add a few more elusive species to the year's list and hopefully take some shots. Thursley lies in the wilds of Surrey, not necessarily a county where you would expect to find an area of intensely rich boggy heathland.

My first trip to Thursley was not a great success. I went in early June, with the self-same Susan who later came with us to Scotland. Despite my badly flawed directions, eventually we were on a familiar-looking straight lane and I glimpsed the sparkle of water through the pine trees. I headed straight for Moat Lake. This is a large, round body of water with thinly vegetated margins, and it tends to attract a slightly different set of Odonata to the more acidic waters out on the heathy expanse of the reserve. It is also home to a motley crew of Mallards, several of which came out of the water, waddling hopefully over at my approach, beady eyes scanning my person for a bag of sandwiches. Some of them looked like standard Mallards, but among them was the odd all-dark or pale-chested bird.

Such ducks regularly confuse beginner birdwatchers, as they don't appear in the field guides. They hang out with Mallards and act just like Mallards, but may be much bigger or smaller, and have all the wrong plumage colours — they may be white or pie-bald, apricot-coloured or dark glossy green all over. The answer is that they *are* Mallards, but of domestic origin. Domestic Mallards, just like domestic cats, dogs, pigeons and goldfish, have been selectively bred to develop all kinds of variants. Unfortunately, some people who decide to get themselves a few domestic ducks for the garden go on to have a change of heart and let the ducks loose at the nearest pond. Either that or the ducks leave of their own accord. So nearly every duck pond with Mallards will have a few oddities like these. Birdwatchers refer to them, rather scathingly, as 'yuck ducks', 'muck ducks' or 'manky Mallards'.

Some of the manky Mallards here had ducklings. There were eight ducklings in all, of similar ages, but they seemed to belong to two different females, judging by the bickering that was going on. Once they realised I had nothing for them they all got back into the pond and swam off disgruntledly to loiter by the bank nearby.

Dragonfly-wise, Moat Pond is home to Downy Emeralds, and also to Brilliant Emeralds. The latter is quite a rare species, with a confusing distribution map — it has a major population centre here in Surrey and surrounding counties, and another in the central Scottish Highlands, with nothing in between. It is speculated that this weird distribution is down to the species making two separate colonisations of Britain, rather than its population becoming fragmented, as there are plenty of patches of suitable habitat in between the two areas where it occurs. It likes somewhat acidic lakes within pine woodland, which is exactly what Moat Pond is.

The Brilliant Emerald is closely related to the Northern Emerald, and not so closely to the Downy Emerald. Its scientific name, *Somatochlora metallica*, means 'metallic green body', which summarises the main way it differs from the other species. All three have very bright green eyes, but while both Downy and Emerald are quite dark-bodied, with a subtle bronzy gloss, the Brilliant has a thorax as bright green as its eyes and a strongly green abdomen too.

Across the water, I could see a number of emeralds whizzing about. It was a little early in the season for Brilliants to be on the wing, and all of the emeralds I could see well seemed to be quite dull and dark of body, suggesting they were Downies. Then I got a binocular lock on one that was heading directly towards me. I could see vivid green eyes and also a vivid green thorax and pretty bright body. It had to be a Brilliant. I watched it get closer, feeling mounting excitement and wondering if Susan wouldn't mind waiting a couple of hours while I tried to take an in-flight photo. Then, as it neared the bank, one of the manky Mallards attacked it, almost caught it, and sent it haring off in the opposite direction. I watched in dismay as it disappeared, and though I waited a while I didn't see it again. That was my punishment, I supposed, for failing to bring some kind of duck food with me.

Leaving Susan enjoying the view from a picnic bench at the lakeshore, I went into the pinewood to try to track down a bird I could hear calling.

It was a fledgling Nuthatch, sitting on a low branch awaiting some attention from its mum, or dad. I located the parent foraging nearby, clambering up and down a snapped-short stump with great agility and collecting a big wodge of crushed flies, spiders and other creepy-crawlies in its sturdy chisel of a bill. In the shady interior of the wood its pretty colours — blue-grey above and pale peach below — were not very clear, until it reached the tip of the stump where it was caught a patch of sunlight. Then it flew to the youngster, which started to beg. The parent unceremoniously stuffed the unappealing-looking mess of squashed insect legs, wings and bodies into the chick's gaping throat and immediately flew off to fetch some more.

We walked out onto the reserve after that. Away from the shelter of the trees, we could feel the considerable force of the wind, and overhead the clouds were moving into position. I didn't hold out high hopes for much Odonata action. At the first mire pond, I set down my rucksack and scanned the water. At first there was no activity at all, until the sun peeped out briefly which inspired a solitary Four-spotted Chaser to take flight. The wind caught it and slammed it rather hard against my rucksack. It clung to the lea side of the bag, gripping the fabric in a state of shock and allowing me to admire its neatly marked wings and great greenish-brown eyes. Then it lifted off again and was immediately blown out of sight.

We followed the boardwalk across the bog. The far-reaching views revealed a flattish, very open landscape that is by no means classically pretty. The heather, yet to flower, was dark and steely looking. Here and there were the snowy tufts of cotton-grass, like little white warning flags positioned around the most boggy and treacherous parts of the heath. A few pine trees broke the skyline, some in tight little crowds on top of raised humps of land, others in neat single or double rows.

We reached the largest pools before the weather had completely failed. In the few moments of proper sunshine that came through as clouds rolled overhead, a few dragonflies appeared, among them one of my target species, the Keeled Skimmer. A delicate light blue male, looking only a little bigger and sturdier than a blue damselfly, settled on a dead sedge stem within easy camera reach. He rested with his wings swept forward fighter-jet style, obscuring his glossy grey-blue eyes when I looked at him side-on. His frons and his head behind the eyes were a contrasting orangey shade, which was matched by his pterostigmas — the coloured cells near the wingtips.

The species name of the Keeled Skimmer, *coerulescens*, comes from the Latin for 'becoming blue', which is beautifully apt. Males begin their adult life as boldly marked black and yellow creatures, but gradually acquire their blue pruinescence over the first few days of existence. The pigmented waxy stuff that provides this colour is actually secreted through the cuticle or 'skin' of the abdomen, and its production is apparently triggered by the dragonfly reaching sexual maturity. The male skimmer therefore slowly becomes more blue as he gets older, but then over time becomes less blue as various stresses of life (especially grappling with females during mating) rub off the pruinescence.

Out over the water I could see several dragonflies flying about, but each one I got in my binoculars was a Four-spotted Chaser, a common species that I knew I would see all over the place. My hopes of finding any other Thursley specialities were fading away. Then a much larger, much faster shape was suddenly belting across my sightline. I quickly drew Susan's attention to the bird and we both watched as it dived steeply downwards, as if about to plunge headlong into the water, and then veer up and away. It was a Hobby, small and agile falcon of heathland and wetland habitats, and arch-enemy of all dragonflies. Although I didn't see exactly what had happened I was sure it had grabbed one of the chasers. It flew away to the north, probably to a nest in one of the big pines.

Susan was thrilled by the Hobby sighting, but less so by the sudden onset of rain. I asked what she would like to do and she said that she would quite like to get out of the rain and find some lunch. When the weather failed to improve over lunchtime so we called it a day.

But I needed to make a return visit to Thursley. There were several more species that I hoped to see there. After our week in Scotland, I contacted Phil Sharp, a local friend who, like me, blogs about his local patch wildlife. He didn't need much persuasion to have an 'off-patch' day at Thursley, as he too had caught the Odonata bug and was keen to see some different species. We were on site by mid-morning with reasonable weather promised.

Thursley is a National Nature Reserve, reflecting its huge importance both for Odonata and for a much wider range of scarce wildlife. It is home to scarce species including Silver-studded Blue butterflies, Dartford Warblers, Woodlarks, insect-eating sundew plants, Raft Spiders, Smooth Snakes and Sand Lizards. Not only that, but it is in itself an example of something rare and threatened — wet heathland, a nationally scarce habitat.

It is a good job that many of the most important bits of Thursley *are* wet, because in 2006 a fire caught hold on the heathland here. The blaze took more than 100 fire-fighters four days to bring under control and damaged more than half of the site. Heathland can burn fast and fiercely, especially after a long, dry summer. The place looked to be in a terrible state and no doubt many animals of all species were killed. But the regeneration has been rapid, and some, such as the Silver-studded Blues, are actually doing better than they were before the fire.

Phil and I started at Moat Pond. Not an emerald to be seen now, but there was another green-eyed something scything low over the water. It was a male Emperor Dragonfly, a really magnificent beast and my first sighting of the year.

The Emperor Dragonfly measures nearly 9cm long, and up to 11cm across its wings. This makes it our largest dragonfly – the female Golden-ringed is slightly longer, but only because of her supersized ovipositor. Its scientific name *Anax imperator* means, pretty much, 'master and commander', the first word being Greek and the second Latin. As well as being huge, it is colourful, the male having a bright green head and thorax and bright blue abdomen, and the female being bright apple green all over. Unlike the hawkers, the abdomen has solid colour rather than spots of colour on a black background, though both sexes have a dark stripe down the centre of the abdomen.

The Emperor is not closely related to our main group of big dragons, the hawkers, although superficially it looks like them. One of the main differences is that the Emperor is nearly always seen over water, while most hawkers will wander well away from the lakes or pools from which they emerged. Another is that the Emperor is faster, fiercer, more agile and more aggressive than the hawkers, and will chase all of them from its territory except the similarly large Southern Hawker.

While by the lake, we explored the shoreline vegetation and found an assortment of peculiar spiders, while further back among the grassy clumps there were several Large Skippers. These are sturdy little butterflies, with broad heads and a distinctive swept-back wing posture. Very bright orange, they were eye-catching when at rest, and in flight looked more like moths with their rapid, buzzing flight, slaloming around the grass stems as they looked for a better basking spot.

We went back through the wood (the Nuthatches were still around, but less showy than before) and then out onto the sandy trail across the heathland. Almost immediately we spotted a Hobby flying overhead, weightless and effortless, swooping and wheeling, describing gently curvaceous lines across the blue sky. Phil, watching through binoculars, saw it seize and then 'process' some unfortunate dragonfly — I missed this as I was struggling with my camera but I did manage a couple of photos after the event, before the Hobby floated away towards Moat Lake.

There are many paths across Thursley. We opted for the one that traverses the wettest part of the heath, so were soon on a wooden boardwalk rather than the sandy trail. Ahead of us, a few Common Lizards were out, basking on the wood. We stopped for photos of the nearest, crouching down to get an eye-to-eye view of the little reptile. It watched us suspiciously, its eyes glittering bright jewels set among in the pretty mosaic tiles of its scaly face. A strongly patterned individual, its colours included black, ash-grey, lemon-yellow, chestnut and dark green. The wood must have been too warm for its toes, because it was resting on its chest and holding its front feet a few millimetres off the ground. We didn't want to disturb it but there was no way round. We came forward slow step after slow step and when we were too close it ran off at speed and darted down a gap between the boards.

The heather here grows tall and tough. It was still not in flower, although the fat cerise blooms of the scarcer bell heather were out, each one looking exactly like a bulbous pink bell with a little frill of flared petal tips around the entrance to the flower's interior. The standard heather had developed tiny pale flower buds but it would be a couple more weeks before it came into full flower and flushed the whole heathland lilac-pink. I noticed a small and darkish dragon buzzing low over the tops of the heather and called Phil's attention to it when it settled.

This was a female Black Darter, *Sympetrum danae*, named for the mother of Perseus in Greek mythology (that's Danae – *sympetrum* means 'likes to sit on rocks'). This little beauty is one of my favourite dragonflies, though it is the males that are the real lookers. This female was pretty enough though, shiny golden yellow with plenty of black on the underneath of her abdomen and her thorax sides. Her eyes had a light brown blob on the top half intersecting with a lemon yellow crescent below, and she hung on to the tip of a heather spring with all six bristly, jet black legs bunched together and her body sticking out almost at right angles.

We were admiring her when we found Mr Black Darter posing in an accessible spot nearby, resting along a rosy-coloured sedge stem. He cut a daintier figure than the female, with a pronounced slim waist halfway down his abdomen, swelling to a club-shaped tip. The segments of his body were of uneven width, creating ridges where they met. He was mostly black but with patches of yellow down the thorax and abdomen sides. His eyes had the same pattern as hers but were darker and greener. Both he and she were in exquisite, perfect condition, gleaming like newly minted coins. The Black Darter is quite a late flyer, the first appearing at the very end of June and peak numbers on the wing in August.

Black Darters are somewhat scarce, especially in southern England, being tied to acidic heathland and moorland with enough standing water for them to breed. It's said that the adults sometimes wander long distances from home but I've only ever met Black Darters in classic Black Darter habitat. The black coloration helps the darter to get warmed up quickly, though it also carries a risk of overheating on really warm days. A Black Darter therefore selects its perching spots to keep its temperature at an optimum level for flight (between 20 to 40°C), and also changes its posture to increase or reduce heat absorption. The male we had found was sitting flat, exposing his whole body to the sunshine on what was a rather cool day for mid-July. However, should things really heat up he would gradually raise up his abdomen, reducing the surface area that was receiving direct sunshine, until his backside was pointed straight upwards in a yogic sun salute.

The path gradually led on across wetter and wetter terrain, and we could see the large shallow pools across to our left. As with my last visit, there were numerous Four-spotted Chasers overflying the water, though few were close enough for a good look. Soon we were surrounded by Keeled Skimmers — mid-July is the peak time for them. Phil had not seen either Black Darters or Keeled Skimmers before and was gratifyingly delighted with both of them, quickly getting his eye in on the blue male skimmers and picking them out from the Common Blue Damselflies that were also numerous here. Once in a while we spotted a bigger, stockier blue skimmer with a black, pruinescene-free bottom — these were male Black-tailed Skimmers. Here, they were vastly outnumbered by the Keeleds, for Keeled is an acid-heath specialist while Black-tailed lives all over the place.

I'd already seen a teneral Black-tailed Skimmer or two on my local patch but this was my first opportunity of the year to admire a mature male in his full glory. These are good-looking dragons, with light blue pruinescence covering all but the black tip of the abdomen, and contrasting pale orange patches down the abdomen sides. As with other skimmers and chasers, the abdomen is rather flattened, wider than it is deep, giving the dragon a robust, utilitarian look. The eyes are a dark, smoky grey-blue.

This dragonfly can be found all over the place in southern England, though is more or less absent north of Yorkshire and is rare in Wales. I remember seeing one in St James' Park in central London, casually basking on a flower-labelling stick in the middle of a bed of ornamental roses. However, it doesn't seem to gather in dense populations — possibly because the males are so violently territorial that they won't tolerate any other males nearby. Of female skimmers there was little evidence — we saw

LEFT:
Vagrant Emperor,
Dungeness,
25 April 2011, 1.05pm

BELOW:
Broad-bodied Chaser,
Pembury,
27 April 2011, 5.51pm

Red-eyed Damselfly, Sevenoaks Wildlife Reserve, 10 May 2011, 9.49am

Beautiful Demoiselle,
Hadlow College,
11 May 2011, 11am

Northern Emeralds,
Coire Loch,
28 June 2011, 1.27pm

Golden-ringed Dragonfly, Bridge of Grudie, 1 July 2011, 1.53pm

Emperor Dragonfly,
Thursley Common,
14 July 2011 12noon

Common Darter,
Rodmell,
29 August 2011, 5.18pm

ABOVE:
Common Clubtail,
Goring,
13 May 2012, 4.03pm

BELOW:
Azure Damselfly,
Sevenoaks Wildlife Reserve,
26 May 2012, 9.05am

ABOVE:
Banded Demoiselles,
Sevenoaks Wildlife Reserve,
26 May 2012, 10.03am

BELOW:
Scarce Chaser,
Strumpshaw Fen,
10 June 2012, 10.50am

Four-spotted Chaser, Strumpshaw Fen, 10 June 2012, 11.40am

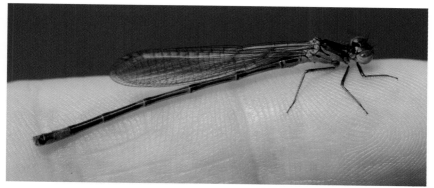

Variable Damselfly, Catfield Fen, 13 June 2012, 3pm

Norfolk Hawker,
Strumpshaw Fen,
10 June 2012, 11.17am

Hairy Hawker,
Catfield Fen,
13 June 2012, 2.06pm

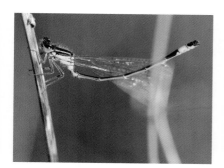

Blue-tailed Damselfly,
Sevenoaks Wildlife Reserve,
20 June 2012, 8.22am

Downy Emerald,
Sevenoaks Wildlife Reserve,
20 June 2012, 8.50am

ABOVE:
Ruddy Darter,
Dungeness,
25 July 2012, 1.19pm

LEFT:
Migrant Hawker,
Rainham Marshes,
7 September 2012,
12.35pm

no Black-tailed and only a few Keeled. This is more of a testament to the females' skill at being discreet than evidence of a real imbalance between the sexes.

Near the far end of the big pools a small bird of prey flew over, very low, giving us a lovely view of its underside and the spread of its mottled wings. I nearly overbalanced into the bog to take photos of it as it passed directly over my head. Not a Hobby this time, this was a female Kestrel, and my photos showed that she sported a metal leg ring. Sadly there wasn't a clear enough view of the ring to see any numbers or code on it so I couldn't find out anything about her life, but I did wonder if, rather than a wild bird, she might be someone's escaped pet, unafraid of people and perhaps even checking us out for any sign of free food. She went on out over the heath and then hovered briefly, perhaps having spotted a lizard. Birds in general had been very much not in evidence so far on our walk so the close encounter was most welcome.

At the far end of the big pools, the path swings right and you stay on a boardwalk that crosses a narrow channel of open water, which is black with peat and so looks much deeper than it is. A wooden railing on one side helps prevent the keen Odonata-phile from falling in and discovering exactly how deep the water is. We had just embarked on this stretch and were a few paces apart when both of us exclaimed excitedly and tried to attract the other's attention. Phil had found a male Emerald Damselfly resting among sedges, while I had just clapped eyes on my very first Small Red Damselfly, also a male, resting on the handrail. Where to look first? I took a quick couple of photos of the Small Red, then went back to check out the Emerald.

There is scope for confusion here, as in Britain we have several species of emerald damselflies *and* also three species of emerald dragonflies. The emerald damsels are sometimes called 'spreadwings', and if this name catches on (which I think it will, though at the same time I'm not quite ready to make the leap myself) then the problem of there being too many emerals will be solved.

This male Emerald Damselfly aka Common Spreadwing was showing off the attributes that inspired both of his names. The wings were held out and proud, at 45 degrees to the body rather than 90 degrees like a dragonfly, revealing a remarkably slim body that reflected deep, metallic emerald green, turning more bronzy near the tail end. A dusting of blue pruinescence just behind the head and on the tail-tip and base, and pretty sky-blue eyes, completed a lovely-looking damselfly. Phil enthused happily about having found one of his favourite species, and we jostled for position to get a clear photo of the little beauty, tricky as he was sitting among a big clump of sedge stems. Luckily we then found a second individual, this one sitting in a much more accessible position on top of a little pine sapling, and the camera-work became a lot easier.

The genus name of the emerald damselflies, *Lestes*, means 'pirate' or 'robber' in Greek. It is a curious name, implying that these damsels steal from other predators — something that, as far as I can find out, they do not do. The species name of the Emerald Damselfly is *sponsa*. Just as puzzling, this is a Latin word meaning 'bride' or 'bridal dress'. So we were looking at a 'robber bride'... and a male one at that. Maybe the blue touch on the male's upper thorax was suggestive of a bridal veil to whoever chose the name. I could see that, just about. Female Emeralds lack the blue, being shiny dark green all over, and are also more thickset in the body.

Although it is by far the commonest and most widespread of the *Lestes* species in Britain, and according to its distribution map should be found virtually everywhere, the Emerald Damselfly still isn't that easy to see. It prefers places just like Thursley, with pools on acid heathland. It occurs in more standard wetlands too, well-vegetated lakes, ponds and creeks, but not all of them — for example, I've never seen one at my local patch, where there are lush lakes aplenty.

The Small Red Damselfly on the other hand is a properly scarce species, and having taken a bushel of photos of the two emeralds I was keen to re-find the male I'd just seen. With Phil's assistance I relocated him – or another exactly like him – now resting along a sedge stem. We had already seen a few Large Reds, but this was a distinctly different creature. Petite and dainty, his long abdomen was bright blood-red from base to tip, his thorax blackish with a faint red gloss. The Large Red has a black tail-tip, and red antehumeral stripes. It is also a bigger and more robust beast than the one I was looking at. This Small Red, by contrast, looked like he would snap in half in a strong breeze.

Peering more closely at the little creature, I saw that he also had red legs, red eyes, red pterostigmas and a red frons – the only black bits were his thorax and the head between the eyes. He was slightly backlit, which highlighted a coating of translucent fuzzy hair over his face and thorax. It was hard to imagine that this sweet, frail insect was actually a consummate predator.

The genus *Ceriagrion* is a large and widespread one, with examples throughout Europe and further afield, including the wonderfully named Suave Citril of Africa. However, here in Britian the Small Red Damselfly is the only representative. As far as I can tell, the prefix *ceri* comes from the Latin word for 'cherry', which seems fitting enough as many *Ceriagrion* species are bright red. *Agrion* is an old name used for many kinds of damselflies. The Small Red's species name, *tenellum*, means 'delicate' or 'tender' in Latin. Clearly I wasn't the first to be struck by the damsel's particularly fragile appearance. In Britain, only in southern heathlands does the water get warm enough for the species' eggs and nymphs to develop properly – most of the world's Small Red Damselflies live around the Mediterranean.

I had hoped to see some female Small Reds, but the only one I did see was intimately entwined with a male so I couldn't get a proper look at her. The females, as with some other damsels, are polymorphic, and there are three distinct forms. Some are coloured the same as males with red abdomens, and are only distinguishable (to an amateur like me, at any rate) from the males by their slightly longer and fatter bodies. Some have completely black abdomens, making them the blackest of all damselflies you'll see in Britain. However, most are half and half, with extensive red at the abdomen base and a little bit of red at the tip, and black in between. These three forms have their own scientific names, albeit unimaginative ones. The reds are *erythrogastrum* ('blood-red stomach'), the black ones are *melanogastrum* ('black stomach') and the intermediate ones are *intermedia* (no translation required).

Small Red Damselflies have a restricted and patchy distribution in Britain, with just five well-spaced population centres. These are the New Forest, the Surrey heaths, the Norfolk Brecklands, the moorland of Exmoor and Dartmoor, and the coastal heaths of north Wales. The best sites have good populations of a rather nice-looking pink flower called marsh St John's wort, although the damsel doesn't need the flower, they just happen to like the same kind of shallow, boggy pools.

We walked on, through a line of smallish pine trees. The day had been remarkably quiet up until this point, with so little birdlife around. You should not gain the impression that Thursley is a bird desert — it isn't, and hosts many interesting species although one of the stars, the lovely Dartford Warbler, does seem to have disappeared. This could have been a result of the 2006 fire, but the very cold winters in 2009–10 and 2010–11 would not have helped either. The Dartford Warbler, a dark little bird with a hectic disposition and a long tail that it holds cocked up like a Wren, is a bird of southern Europe really. It reaches its northern limits in England, where it lives on warm open heathlands, especially in the southwest. Unlike most warblers it does not migrate, and so can really struggle when there is a long freeze-up. Depending on weather, I would not be at all surprised if 'Darties' were to return to Thursley but if there were any there in 2011 they were keeping a very low profile.

The other special breeding birds of the heath, the Woodlarks, Curlews and Stonechats, probably were around but high summer is when they are at their quietest and most discreet. Another speciality, the Nightjar, is active only after dusk, and while I have watched them hawking moths over the Moat Pond on July evenings in previous years, today we didn't have the time to wait for them.

The line of pines, though, was suddenly alive with bird sounds and we spent several minutes trying to track down the sources of the various squeaks and churrs we could hear. Every so often a shadowy little bird shape would appear and then vanish among the pines' thick foliage — we could only register something slim and large-headed with a pointy length of tail. Patience paid off in the end and we managed some clear views of a family of Whitethroats — dapper little warblers that can be very common in open, scrubby countryside. The youngsters were only recently out of the nest, with that rather loose-feathered, fuzzy look about them, and dark eyes. Their mother, more alert and very difficult to see, had lighter-coloured eyes and the careworn plumage of a bird that has been run ragged after her babies for the last few weeks.

We were quite close to completing our circular walk, but had to reconsider our options when we found that a section on the return part of the loop was waterlogged, where the old boardwalk had given up the ghost and most of it had come loose and floated off in random directions. We could leap the gaps between the boards that remained, or we could retrace our steps. Phil was game for the leaping option but I was less enthusiastic. I'm not naturally good at jumping or balancing, especially not with an expensive lump of Nikon swinging around my neck. Phil kindly agreed we could go back the way we came.

We stopped on the way a few times. The first was to enjoy a rather splendid Four-spotted Chaser that had chosen as its lookout perch a photogenic dead stick, poking up out of a mire pond. Phil told me that this was his favourite dragonfly species, and I could see why. It isn't colourful but it has a lot of style in its understated way, and a self-confident panache in the way it conducts itself. We both took numerous photos of the chaser as it sat regally on its perch, its pointed black rear thrust out at a right angle to the stick. Then it dashed off, snared a big, fat fly in the basket of its cupped legs, and returned to its perch to consume the feast. The fly was considerably larger than the chaser's own head, so the effect was rather as if a human had settled down to enjoy an entire roast turkey in one go.

The Four-spotted Chaser takes its name from the dark patches of colour at the mid-point of its wings, along the leading edges. Its scientific name, *Libellula quadrimaculata*, means almost the same as its English name – 'dragonfly with four spots'. All Odonata have a slight notch in the costa (the main vein that runs along the wing edge) here, where a major cross-vein branches off the costa, and it is this notch or node that is marked with a blob of pigment in the Four-spotted Chaser. It also has a bit of extra dark colour around the pterostigmas, a large dark triangle at the base of the hindwings, and a strong yellowish flush at all wing-bases. In short, it has the most decorated wings of all the breeding British dragonflies, which compensates for its dull grey-green body, marked with yellow patches down the sides. There is actually a rare variant of the Four-spotted Chaser with a complete dark tip or band across each wing, but I've yet to track one of these down.

I have tried to find out whether this chaser's spots are used for display, or whether (like the dark pterostigmas of most species) they provide extra strength and assist flight stability. I couldn't track down a firm answer, but because the spots are present in both sexes, I am inclined to think the latter theory is correct. This chaser is unusual in that the males and females look very similar, with just a slightly fatter (but still strongly tapered) abdomen and different-shaped anal appendages picking out the females.

The Four-spotted Chaser is a very common, very widespread dragonfly, because of the nymph's tolerance for all kinds of habitats (even including slightly salty water near the coast), and its ability to quickly colonise new wetlands. It lives right across Europe and also in North America, and is a noted traveller, undergoing mass movements every few years (probably in response to overpopulation and overcrowding). When this happens, the moving insects gather and travel in colossal, tightly packed swarms, sometimes comprising more than two billion insects. This must be a staggering sight – and one that is very rarely seen in Britain. I found a few images online of cornfields full of resting chasers and skies full of them on the move, but can only imagine what it would be like to see – and hear – a swarm in real life.

At Thursley there were probably not two billion Four-spotted Chasers but there were certainly plenty. As we watched the frenetic action over the water, one chaser came to a sorry end when it was seized in its low, patrolling flight by a Mallard duckling, of all things, which had been squelching around in the shallows. The duckling managed to slam the dragonfly down into the muddy water, and then after much struggling manoeuvred its prey, plus a good helping of boggy mud, into a bill that was not at all well adapted for the task. It's a tough life being a dragonfly – fierce predators though they are, there is always a bigger, if not necessarily badder, predator out there.

One such predator was at work over a small pond on the other side of the path. Phil pointed it out – a male Emperor Dragonfly, much closer than the one we had seen at the start of our walk over Moat Pond. We stopped yet again to watch him in action, and both of us had a go at getting some flight photos. Cue many sighs of frustration – the Emperor was in full hunting mode and therefore was jinking around, changing speed and generally being impossible to track. We gave up and just watched, as he sailed and zigzagged and swooped after midges so small that he could easily scoff them in flight without missing a wingbeat. With his great pale greenish eyes, lidless, pupil-less, expressionless, so unlike our idea of proper eyes, he looked more machine than animal, and he carved a path of perfect efficiency up and down his patch with no sign of tiring.

I knew that if the Emperor caught some larger prey he might land, and after we'd been watching for about 10 minutes this is what happened. I didn't see what he had caught but he made an elegant landing on a sedge nearby and Phil and I hurried over, cameras-first. Phil was stopped in his tracks by a ringing phone, while I went on and began to take pictures of the Emperor enjoying his feast.

The prey, it turned out, was a two-course meal — the dragon had caught a pair of Common Blue Damselflies in tandem. By the time I started to take photos he had already consumed the head of the male, and the rest of the unfortunate damsel's banded blue body was slowly disappearing, millimetre by millimetre, between the Emperor's methodically working mouthparts. The male damsel's back end had lost its grip on the female, but she could not escape, as the Emperor had casually pinned her against the sedge with one foot, apparently keeping her for later while he scrunched up her partner. I felt glad that damselflies have limited intelligence, as surely this must be about as horrifying a fate as anyone could imagine.

Seeing them side-by-side, the size difference between dragon and damsel was astounding. The female damsel's body was barely longer than the top joint of the Emperor's leg, her little dumbbell head a tenth the size of one of the Emperor's eyes. I was making various exclamations at the spectacle, while Phil, who had come over but was still on the phone, was practically jumping up and down with frustration at missing this remarkable photographic opportunity and doing all he could to politely end a conversation with someone who had obviously called in hope of a really long chat. In the end, in desperation he took a few photos one-handed — difficult with a DSLR plus telephoto lens as the heavy lens really needs some support — but they turned out well, all things considered.

So while dragonflies have it tough, the damsels have it much tougher, and in this real-life world of deadly monsters and their frail, helpless victims there are no knights on white chargers to step in when a dragon captures a damsel. However, things worked out all right for this particular damsel in the end. The Emperor finished eating the male damsel and then simply flew off, releasing the unhurt female to live another day. Maybe he hadn't even realised she was there.

We kept going, leaving the larger lakes behind. But there were still plenty of small mire pools to walk past, and on one of these we found a female Emperor Dragonfly, busily ovipositing. She was facing us, sitting precariously on a wodge of floating leaves, with her wings spread in a bi-plane position and uptilted, ready to lift off if necessary, abdomen gracefully curled down into the water. At this moment, danger in the form of a predatory fish or bird could come from above or below — and she would be hard-pressed to escape.

It was interesting to see this other, more vulnerable side of the mighty Emperor. Adult dragonflies are masters of their aerial environment, just as their nymphs are a force to be reckoned with in the underwater world. But when the fully developed nymph crawls out of the water for its winged adult self to emerge, and when the adult female returns to the water to lay her eggs — these are moments of real weakness in a dragonfly's life.

When it comes down to it, dragonflies don't have a lot of ways to defend themselves from predators. They are, when flying, quick enough to evade most other aerial hunters, but when at rest and slowed down because of cold, they could fall prey to many bird species. If they rest too near the ground they may be eaten by frogs and also mammals like Stoats and even Hedgehogs. I've even heard of dragonflies being killed by Hornets. The dragonfly has no sting, no powerful kick, very little to protect itself.

However, dragonflies can, apparently, bite. Well, we know that they can scrunch up other insects, but I had always believed that they couldn't and wouldn't bite in any significant way. But it turns out I was wrong. I have been reliably informed that they can bite a person hard enough for it to hurt, so presumably could also do the same to another animal that was trying to eat them, though whether it would be enough to deter the predator is another matter. I haven't personally received a dragonfly bite, and from what I've heard it's much less painful an experience than a chomp from a horsefly or a sting from a wasp, but it is enough to make you not want a repeat performance. This should not deter you from spending time with dragonflies though, it is not as though they will chase you down and attach themselves to your arm. However, if you decide to try to pick one up, you'd be better off encouraging it to step onto your hand of its own accord rather than trying to wrap your fingers around it — that's when it might give you a nip.

Phil and I carried on along the boardwalk and were soon back at the car park. We were very pleased with our day of dragonfly watching but there was one more non-Odonata species that Phil was keen to see — the Silver-studded Blue butterfly. This is a rare butterfly, only found on certain heathlands, and Phil had never seen it before. I had, but not nearly often enough, so was keen to get acquainted with it again.

The first time I ever saw Silver-studded Blues was also the first time I visited Thursley, back in 1995. I went with my boyfriend of the time in mid-June, arriving at about 9am. We had barely set foot in the reserve before we found a Dartford Warbler, singing away from the tip of a tall heather clump, flirting his long tail, flaunting his outrageous bright red eyerings and generally making up for the fact that his dry, scratchy little song was a bit rubbish. A little further on we saw our first Silver-studded Blue, a male, cracking open his wings just enough for the shining blue inner surface to show. As things warmed up, more and more of the blues appeared. They had been there all along, but were almost invisible with their wings closed. As the sun's warmth began to kick in they all started to open their wings, and it was as though the heath was bursting into bloom with little blue flowers opening up everywhere we looked.

Today there was no such spectacle. By mid-July the flight season of this lovely insect is nearly over. But Phil and I took a path into a drier part of the heath, scoured the heather, and eventually picked up a rather ragged brown female dancing along. She settled and gave us a view of her silver studs — a row of spots on the outer edge of her underwings that sparkled in the light. We then found another female, and a very faded male. This time next week there would probably be none left at all, but instead there would be a new batch of tiny, green, woodlouse-shaped caterpillars nibbling their way through the freshest young heather shoots, on their way to becoming next year's beautiful butterflies.

CHAPTER 15

Gomphus vulgatissimus
The Common Clubtail

ᘓ ᘔ ᘓ

S pring 2012 and it was time to tackle last year's nemesis. Our 2011 dragon season got off to a great start with the Vagrant Emperor at Dungeness, but missing the Common Clubtail in June was a huge disappointment. Still, it was a new year and I felt better prepared this time, having done lots of research and checked for other people's sightings of the species every day since the start of April. It was 12 May when the first clubtails were reported on the Thames, and with great good fortune the next day dawned reasonably clear and sunny, and it was a Sunday. We met up in the afternoon and began the long drive westwards.

I spent the journey quietly fretting. Even though our original vague idea to see all the British Odonata in one year had long been abandoned, I particularly wanted to see this species, and felt that we needed a real stroke of luck to be successful. My feelings for *Gomphus vulgatissimus* were worryingly intense. I did, though, immensely enjoy saying or even just thinking its scientific name — what a big, delicious mouthful of words. I found out that *Gomphus* is derived from a Greek word meaning a plug or wedge-shaped nail, accurately descriptive of the dragon's shape, while *vulgatissimus* is an exaggerated form of *vulgus*. This is the same word from which we get 'vulgar' but in scientific name-speak *vulgus* just means 'common', and not that the animal in question is in any way uncouth. So *vulgatissimus* means 'very or extremely common'. Like many so-called common animals, though, the Common Clubtail is in fact not at all common in Britain.

It is a dragonfly of languorous and unspoilt lowland rivers with soft, silty bottoms, which aren't easy to come by in Britain. Seven rivers support populations of the species: they are the Dee, Severn, Wye, Tywi and Teifi originating in Wales, and the Thames and Arun of southern England. Its scarcity makes it a species of some conservation concern, although on a global scale it is in no trouble at all (for now at least) with a massive European range stretching eastward to Russia.

The genus *Gomphus* is very large, with representatives around the world. Many of the 50 or so species are denizens of North America, and rejoice in such names as Cobra Clubtail, Pronghorn Clubtail and Mustached Clubtail. All share the same striking body shape, with the abdomen flaring out to a heavy, thickened tail.

Once we left the motorway somewhere near Reading, the challenge of finding a good place to park near the river began. I had in mind the quiet tree-shrouded patch of gravel where we left the car last year, but we did attempt to find a better spot first, which would give quicker access to the best bit of riverside and avoid the cow-filled field that held us up last year. So we turned riverwards a little earlier along the route, in Pangbourne, only to find ourselves heading for a toll bridge over the Thames where we grudgingly paid up and then up into the hills with no option to park anywhere near the river.

There followed a short jaunt through the rather beautiful rolling Thames-side country. The scenery here is lovely in a very English way, with the contrasting greens of hillside, hedgerow and thicket, and the scattered country houses gleam with prosperity. Through a mostly blue sky, the wide-winged shapes of Common Buzzards and the fork-tailed outlines of Red Kites drifted serenely by. Finally we descended the long slope into Goring, back to the spot where we'd parked last year.

The first stretch of riverside path went through a glade of trees, their roots exposed by the waterside where the river had slowly but surely carried away the supporting soil, to dump it elsewhere. Out on the water a slow trickle of boats went by, from canoes to hefty cruisers, everyone enjoying a rare sunny day in what was shaping up to be a gloomy spring. Soon we were out of the trees and braving the cow field. A number of the cows crowded close to the gate, and observed us with mild interest as we passed them. I don't mind admitting that cows make me nervous. So much bone and burly muscle, so little fear, so many good reasons to not look kindly upon *Homo sapiens*. But of course the cows weren't interested in us, and we left them behind as we neared the crucial railway bridge where all the *Gomphus*-spotters go for their best chance of meeting a clubtail.

At the bridge, it was time to take stock. Although the weather was still reasonably encouraging, we'd seen precisely zero dragons or damsels on the way here. Rob opted to stay put on the riverbank while I decided to actively search this stretch of bankside vegetation.

Sometimes the dragonflies and damselflies have come easily to us — go to the right place at the right time and there they are. With the Common Clubtail, I never expected things to be simple, not even when we tried the first time, and today my hopes were very low. Knowing they'd been seen at this very spot yesterday should have given me encouragement, but the place seemed to be nearly devoid of all insect life, let alone dragonflies. However, there was one encouraging sign.

Downriver, I noticed a long-winged bird in agile, twisty flight quite high over the water. Rob spotted it at the same time and called it — 'Hobby'. Impressed with his fast-improving bird watching skills, I had a look through my binoculars (as it was too far off for worthwhile photos). It was indeed a Hobby, a falcon noted for its appetite for dragonflies. Did this bode well or ill for our search for *Gomphus vulgatissimus*? What if I saw this Hobby casually catch and dismantle a clubtail-shaped prey item in mid-air? I decided that this was an eventuality I simply couldn't afford to worry about. There had been a big arrival of Hobbies into Britain over the last couple of weeks, and by all accounts most of them were filling up on St Mark's Flies — jet black, fat and rather dopey insects that flew with dangling legs. Traditionally the flies appeared in profusion around St Mark's Day – 25 April – but like many other things this year they were a little behind schedule.

The Hobby was a delight to watch. A streamlined little fighter jet of a bird with a fierce, moustachioed face, it looped and rose and dived, skimming along the treetops by the far bank. Now and then it would bring its talons up to its bill to eat whatever it had caught, showing the full length of its slim yellow legs in their feathery russet trousers. Once a rather scarce British bird, the Hobby has been increasing and spreading more widely in the last few decades. It is a summer migrant, with most birds hitting the British coast in late April and early May. Concentrations then form at the same sorts of places (mainly insect-rich wetlands) where Swallows, Swifts and martins gather before they breed. I'd read recently that the wetland nature reserve Stodmarsh in Kent was hosting more than 50 Hobbies, all hunting over the reedbeds together on sunny days – what an amazing sight that must be.

The Hobby finally retreated further downstream and I set about dragon-hunting in earnest, poking around in the lush long grass along the riverside. Quite soon I spotted a long-bodied insect flitting among the leaves of a bush and felt a little surge of excitement, but realised more or less straight away by its shape and colours that it was a female Banded Demoiselle. She landed on an arching grass stem and I leaned in for a better look. The season had not long begun for this species and this individual was looking pristine, all polished emerald body and green-gold wings. Yet there were no male Bandeds around to appreciate her newly minted beauty. The frenzy of mating and egg-laying activity among the Bandeds that we'd seen here last year wasn't going to be repeated today.

In between scrutinising the vegetation I was looking into the sky. For the locals, Red Kites are nothing special, or maybe I should say nothing exceptional, any more. I know people living in Reading and thereabouts who see kites overhead every day, kites in the garden looking for scraps, kites as constant companions in their daily lives. But these fabulous birds of prey have yet to appear in good numbers in my local area so I was eager to enjoy them.

A dragonfly has absolute mastery of the air and so does a Red Kite, but in very different ways. While the dragon moves with sharp turns executed with split-second precision – forwards, backwards, up, down – the kite just rides on the air like a gull rides on the choppy sea, lifting and dipping on the air currents on its huge spread of wing without a trace of apparent effort. Yet when it needs to – in dispute with another bird or when diving after some tasty morsel on the ground, it can manoeuvre with stunning agility. The long, deeply forked tail that is its most striking feature is twisted through 90 degrees either side as it steers its course.

A few Red Kites did waft over, though they were mostly very high. Then I spotted the Hobby again, much closer this time, and enjoyed a real show as it went chasing and plunging after flies, up and down the river in front of me. I took photos, of course, though only in a few cases did I catch it properly, sharp with its well-patterned breast and belly showing clearly against the confused dark backdrop of trees on the opposite bank. One of the shots shows it full face to the camera in dramatic cruciform shape, bringing some tiny just-caught victim up to its bill, its clasped feet drawn up under its bowed head as if in prayer. A family out with their dog stopped to admire it and asked me what it was – the father, who clearly knew something about birds (enough to know he wasn't looking at yet another Red Kite, at least) was delighted to discover it was a Hobby.

Quite a few people had gone by after we'd been on site for an hour, though I was a bit discouraged that none of them seemed to be fellow dragonfly-searchers. I was as sure as I could be that we were in *the* spot, and certain that the quarry had been seen here just the day before, it was a sunny Sunday afternoon, so where was everyone else? What did they know that I didn't? Or was it simply the case that dragonfly-searchers just aren't as common as I think they should be? I'd have been more encouraged if there'd been just one other person, but it was just Rob and I, and time was wearing on.

I moved off to search the small field south of the bridge. This was full of flowers and fast-growing grass — it looked a suitable hiding place for any newly emerged insect that was feeling vulnerable. I knew that if we did find a *Gomphus*, that is what it would be — a brand new emergee, eager to get away from the riverside and into a more sheltered area where it would grow and mature for some days before it reached its prime and felt the need to mate. Then it would return to the riverside in its glorious maturity and look for others of its kind. But the fresh *Gomphus* is a fragile thing, and can only fly short distances on its soft new wings, which is why looking for teneral individuals near the river can pay dividends.

I made my way to the far corner of the field, along a thin trail through the grass, and here my adrenaline levels briefly spiked again when I noticed a second female Banded Demoiselle. Then, turning back towards the main path, I saw another Odonata flitter along and land close to me. It was the length of a demoiselle but much chunkier. It was black with bright yellow markings. Its abdomen started out with a narrow waist, but swelled to a broad tail. It was a Common Clubtail, and finally my adrenal glands had something meaningful to do.

I crouched down and got photographing. The clubtail seemed quite happy where it was, squatting in the cup formed by a young Stinging Nettle leaf-head, and I was taking frame-filling shots. It rested with its wings drawn back and pressed loosely together, damselfly style, giving an unimpeded view of the rest of it. The only parts of it that were not black or yellow were its eyes — a pale milky green-grey with a diffuse pseudopupil at the centre. A close look at its head revealed a form very different to that of typical dragons and somewhere in between dragon and damsel. The eyes were relatively small and did not touch at the front, meaning that above its frons it actually had what amounted to a forehead, coloured with a broad band of black and a narrower one of yellow above. This gap between the eyes was surmounted by a short crest of upright black hairs, like a back-combed fringe.

I called Rob over. He was reluctant to come, having plenty to carry and apparently thinking I'd found a ladybird, wasp or something similarly off-plan, but when I'd communicated through shouts and excited hand gestures that no, this was *it*, he gathered up all his stuff and came to where I was, following my directions and picking an ever more tentative course towards me until he spotted the dragon, still reclining attractively on its nettle perch. He busied himself with macro lens, flashgun and whatnot, while I waited, essentially trapped because the only feasible way back to the main path would have forced me to step very near, if not actually on, our precious clubtail.

The dragon's patience outlasted mine in the end and I went back past it, stepping with extreme care. It wasn't disturbed and stayed put for a few more shots, but when it did finally make another short flight 10 minutes later, moving a few metres further away from the river, it settled somewhere we couldn't get to so we decided to leave it be.

With a long drive home, we decided to go, very happy with the day's work. There was one last tricky moment — the cows in the cow field had by now moved right up to the gate and completely blocked it. In a not particularly tense Mexican stand-off we waited for a while to see if they would move. They didn't. Most peacefully carried on cropping the grass, while a couple just gazed at us, emanating a total lack of concern with every slow breath. We ended up walking across to the far side of the field, where we found a gate to a small trail that led back to the waiting car.

CHAPTER 16

Calopteryx splendens
The Banded Demoiselle

ꙮꙮ

After one very sunny and absurdly hot week in April, spring-like weather in 2012 was in very short supply. It seemed that the very moment a hosepipe ban was announced for several regions including the south-east, we began to experience weather that belonged in a tempestuous late autumn — cold, blowy and punctuated with downpours.

The wildlife suffered. A snapshot of what was going on was provided by the growing numbers of people who have a nestbox with fitted webcam in their gardens. Instead of providing charming scenes of growing Blue Tits and Great Tits, the webcams were recording starvation and death, as the adult birds struggled and failed to find enough insects to keep even one or two of their chicks alive. Webcams looking into the nests of birds of prey like Peregrine Falcons and Ospreys told a similar dismal story. At Dyfi in Wales, a rainstorm that just wouldn't end kept a mother Osprey flattened on her nest for a day and a night, while sheltered beneath her the young chicks were safe and warm but slowly starving. By the time the rain stopped and the parents could hunt fish again, just one chick was still alive, and only barely, too weak to beg its parents for food. Only its removal from the nest for half an hour of feeding from the reserve wardens saved the youngster's life. *Springwatch* viewers' hearts melted up and down the country as they watched footage of the workers patiently coaxing the tiny, reptilian nestling to accept miniscule scraps of food.

Floods swept away waterfowl nests and swamped newly hatched wader chicks. Rising rivers ate up muddy banks where Kingfishers and Sand

Martins had made their burrows. Swifts and Swallows had to delay the start of their nesting season as they couldn't find enough food to reach breeding condition after migration. It was no spring to be a bird, and it seemed not much of one to be an insect either. The cold air made flight difficult, so for flying insects needing to travel to find food and mates, opportunities to do anything but sit in sheltered vegetation waiting it out were limited. An insect can last a surprisingly long time in a motionless, zero-energy state, but not forever.

For the dragons and damsels, most of them still in their nymph form, living underwater and waiting for the right time to emerge, maybe the bad weather was not so disastrous. Or maybe it was — flooding stirs up the sediment on lake and riverbeds, making it hard for the nymphs to find food and perhaps washing them away from the good feeding grounds. Really turbulent conditions will just crush them. On the other hand, extra rainfall means more standing water, which could work out well for those that do survive and emerge later in the year, allowing colonies to spread and grow.

Non-stop bad weather is bad news for dragonfly-watchers too, of course. Opportunities to get out and see Odonata were few and far between. One Saturday towards the end of May there were a few sunny hours predicted first thing, not long enough to make a decent trip anywhere very far away, but it was a chance to see how the local species were doing. I packed up at dawn and headed for the local patch, Sevenoaks Wildlife Reserve, by myself.

The reserve's wildlife garden was my first port of call. An inspiring little patch, this, with lush borders of lovely native flowers like Foxgloves, a bird feeding station with a simple screen allowing close views, and a small, well-vegetated pond in the centre, the sort of pond that would work well in even a quite modest-sized garden. It was already quite warm and there was much damselfly activity. Many of these could have come from the large lake nearby but others would have emerged from the little pond.

I patrolled the borders, looking for damsels perched at a good height for photos. There were plenty of them — adults and tenerals all together, though they were lively in the warmth and didn't allow me to get too close. There's a knack to this, though, and as I worked I started to remember how to do it. I think of it as the 'looming technique'.

Dragons and damsels are very alert to sideways movement. If you, or a part of you, move across their line of sight, even quite slowly, they will often take flight. So the trick is to stand back, directly in front of the resting insect, and then move forwards as smoothly as you can, so that from your target's point of view, you seem not to move but to expand. Have the camera raised and ready as you do this. Take care not to let your shadow fall on your subject, this is sometimes enough to convince them to fly. And use a long lens if you can, so you don't have to get that close for a frame-filling view.

Most of the damsels here were Azures, the commonest of the diverse *Coenagrion* genus in Britain. The electric-blue males were most eyecatching, and I was particularly pleased to find one resting on an unopened poppy flower-bud. The bud's perfect plumpness and fuzzy texture, suspended from a U-turned stem, made a lovely contrast to the damsel's straight, slim lines and enamelled smoothness. Not to be outdone, a female Azure settled nearby in the pretty cup of an Ox-eye Daisy, her black-and-grey eyes peering between the shaggy fringes of white petals. Besides Azures there were also Blue-tails and a handful of Large Reds and Common Blues. They were resting and hunting, without much sign of urgency. It was still early in the day — damsels tend to take it easy until midday, when if the weather is still good the mature individuals will head for some suitable water and search for mates.

Some of those that I wanted to photograph were sitting low down, which meant I too had to get low to take the photos I wanted. Having reached an age (and perhaps also weight) where I don't feel quite as confident in my knees as I used to, this gave me brief pause for thought. But

I did have an advantage — more than ten years of aikido training. A key component of the style of aikido I studied was *suwari waza* — techniques executed while kneeling. Japanese warriors spent much of their social lives in a kneeling or *seiza* position, and moving about while kneeling with the toes curled under to permit quick 'steps' (*shikko* walking). Therefore they needed to be able to repel attackers just as effectively from down there as when on their feet. I can't say that *suwari waza* is my favourite part of aikido training, but the ability to move smoothly, quickly and easily around on my knees has proven a real boon for wildlife photography. I *shikko*ed about, taking photos and getting my trouser knees soaked with dew, until I felt I'd more or less exhausted all photographic possibilities here, and left the damsels in peace.

In just a couple of weeks the plants had gone berserk, piling on great length of stem and density of leaf as they sucked up the abundant rain that had fallen. The familiar outlines of the trails into the heart of the reserve were softened, corners tightened and views hidden. Blackcaps and Garden Warblers sang from the trees and I didn't pause to try to see them, it would be a thankless task. I did pause long enough to look over East Lake and note that the Mute Swan pair there were shepherding three small cygnets along between the gravelly islands — breeding success there at least. I kept going, heading for Long Lake, the best of the lakes for Odonata.

Here, though, things were disappointingly quiet. By scanning the lily pads I eventually picked up a few Red-eyed Damselflies, the males stunning with their black and blue bodies and scarlet eyes that seem lit from within. Only one individual, bucking the trend by sitting on a reed by the bank, was near enough for a photo. Sadly I couldn't see any Downy Emeralds at all.

I made a thorough search anyway. The lake is popular with fishermen and there are numerous narrow, somewhat treacherous paths over muddy ground to the shore, every one of which I went down for a different perspective on the water. The one concrete pathway was temporarily

closed, a strip of red and white plastic tied
between two metal posts forming a barrier
to discourage the conscientious. From the
next pathway the reason for the closure was obvi-
ous – the pair of Mute Swans that nests on this lake had opted this year to
build their nest right beside the concrete path. Mutes are famously fierce
in defence of their nests – anyone breaching the barricade would have
been disciplined severely, swan-style.

I watched the swans from the safety of the next path. The male was
sitting on the half-finished nest, a soggy-looking pile of half-drowned
vegetation, and from there stretched his elegant neck forth to the surrounding water and grabbed more floating leaves and reeds, adding them
to the structure. The female was gently bobbing about nearby, gathering
nesting material a little more energetically. This pair is one of four that I
know of on the reserve, and despite the lushness of their chosen lake I've
not known them to have great success at breeding time. The two pairs on
the much larger East Lake seem to do better. Even with such a formidable parenting team looking out for them, cygnets are vulnerable to many
predators, from gulls, Grey Herons and Foxes to Pike.

Closer to hand, I noticed a couple of female Banded Demoiselles idly
flying among the waterside vegetation. The light was gorgeous in the lake-
wards direction so I was pleased when one of the little beauties settled on
an elegantly arched reed leaf. Whether it was the light or an actual variation, I don't know, but she looked much more golden than green, the
tracery of her wings bearing the same strong yellowy tone as her body. The
white 'pseudo-pterostigmas' on her wings were strikingly bright.

Anyone becoming interested in dragonfly and damselfly identification needs to be familiar with a number of curious new anatomical terms,
such as 'antehumeral stripes', 'frons' and 'pronotum'. Another to learn is
'pterostigma', which means literally 'wing-mark'. It is a cell of wing membrane, framed by veinwork, on the leading edge of the forewing, near the

tip. It stands out from the rest of the wing because of its colour — usually it's dark against an otherwise clear wing.

Fascinatingly, the tiny and insignificant-looking pterostigma has a crucial role in the dynamics of flight. It is thicker than the rest of the wing and so improves stability during gliding flight by dampening vibrations. Its presence enables a dragon to glide up to 25 per cent more quickly than it could without it. As far as identification goes, the shape and colour of the pterostigma is sometimes a key trait to separate very similar species. The Scarce Emerald Damselfly, for example, has a shorter and fatter one than its relative the Emerald Damselfly.

So that's a pterostigma, but what about a 'pseudo-pterostigma'? Damselflies of the genus *Calopteryx* don't have conventional pterostigmas. Why not? I've tried to find out but without success, though maybe it has something to do with the fact that large areas of the wings in these damsels are heavily pigmented anyway. Female *Calopteryx* damsels do have a prominent white cell on their forewings in the position you'd expect to find a pterostigma. That it's not a true pterostigma is revealed because the cell is crossed by venation. I've yet to discover the supposed function of the pseudo-pterostigma — presumably as it doesn't have the structure of the real thing it doesn't do anything for the insect's gliding speed. Maybe it is ornamental, helping males spot the females among the greenness of their preferred resting spots. Certainly female Banded and Beautiful Demoiselles share the males' habit of wing flicking when at rest, which does draw attention to those little white blobs.

This demoiselle was a dream model. From my perspective her reed-leaf perch was set against the water, which provided a very blurry soft backdrop of dappled blue, highlighting the subject nicely. My happiness as I exploited this photographic opportunity was increased considerably when a second female Banded alighted on the leaf, facing the same way as the first. The two shared their perch quite cordially, one on the upward curve and one on the downward. Now and then one or the other would take off and expertly bring down some passing gnat or midge, and then return to the reed leaf to daintily consume the most succulent bits of her kill. With no sign of any Banded males around yet, the females were apparently enjoying a bit of 'me-time', and building strength and resources for the inevitable exertions of courtship and mating that were to come. Although the species name *splendens* was probably originally inspired by the more colourful males, these two girls looked every bit as splendid to me.

I went on after this to explore the little patch of meadow that adjoins Long Lake at the far end. This is a happy hunting ground for insects, and is a favoured nursery for immature damselflies that are not ready to join the hurly-burly of the courting adults over the water. As I stepped slowly through the dewy grass these teneral damsels flew up in swarms ahead of me, backlit and bright-winged. There were many other insects to see as well and I pointed the camera at some of these.

First was the most enchanting small moth *Pyrausta aurata*, which had forewings of an almost indescribably pretty colour — somewhere between burgundy and violet, decorated with small and large spots of bright gold and with a silvery sheen overall. It rested with these wings half open, revealing hindwings of a deeper and browner tone, with a broad band of gold across them. Both pairs of wings are bordered with a soft fringe of creamy fuzz. Just 2cm across at full span, this fabulous little moth doesn't seem to have an official English name though is sometimes called 'Mint Moth' in honour of the plant on which it lays its eggs. My Mint Moth was scrambling up the square-cut sticky stem of some Goose Grass when I grabbed its photo.

I found a bigger, equally attractive moth poised at the tip of a grass blade. This one was mostly white but with two bands of complex grey-brown marbling across its wings, one just below its head and the other halfway down. I knew from its wide, frail-looking wings and slim body that it was probably a representative of the enormous family Geometridae, which includes the 'carpets', 'pugs', 'rivulets' and 'pinions' among many others, but I could get no further with its identification and so took photos to refer to when I was in front of my PC. Back home later it took a long trawl through the UK Moths website images (ukmoths.org.uk) before I felt reasonably confident in calling it a Silver-ground Carpet, *Xanthorhoe montanata*. I felt the stirrings of a need to get more familiar with the moths of Britain — for their wonderful nomenclature if nothing else.

The next miniature inhabitant of the meadow to draw my attention was a beetle, laboriously wobbling along a grass blade that was swaying in the breeze and threatening to unseat the little creature. Sharp photos were therefore difficult to achieve but I managed a couple. This was a Red-tipped Flower Beetle, also known as Common Malachite Beetle, scientific name *Malachius bipustulatus*. Without looking it up I think that means 'malachitey thing with two angry-looking spots', an accurate description. The beetle is small and rectangular, and is a glossy metallic green with a bright red spot on the tip of each wing-case, as if it had recently sat down in a beetle-sized pot of scarlet paint. This pretty colour scheme, together with its big dark eyes and engagingly long and hyperactive antennae, make it a very appealing little insect.

I made my way back past West Lake, which is large and deep and not generally that inviting for Odonata. However, there were a few damsels at the water's edge, including a teneral so fresh it still looked damp all over. I picked it up easily and raised it to eye-level for a close look, but I couldn't identify it, which was a bit of a blow considering how far my identification skills have progressed in the last two years. It was a light tawny brown, a little darker on the thorax top but with no clear antehumeral stripes, and the thorax sides showed no discernible markings that I recognised. I gave up trying to work out what it was and just enjoyed scrutinising it. Then I tried to take its photo. This proved difficult in the extreme, as the damsel was on my left hand and I really need both hands to manage my camera plus 180mm macro lens. I also need, ideally, to get the camera further than an arm's length away from my subject: at this distance, I could only fit half of the 2.5cm damsel in the camera's frame. Still, I gave it a go. Of the photos, one came out quite nicely, a slightly oblique head-and-shoulders portrait at the little creature, its left eye in perfect focus and everything else fading away. As alien as that spherical eye looks, it still seems to me to hold an anxious expression. My very best guess as to its identity is an immature Blue-tailed Damselfly.

There were no more Odonata to see that day, so I switched from macro to telephoto lens and turned my attention to the birds. Finding an active Great Spotted Woodpecker nest hole in a dead stump, I waited out of view under a nearby tree and when the adult male came to the nest I pointed my camera up through the leaves and took some green-veiled shots of him braced there, bill packed with insects for the clamouring chicks inside. The food delivery was over in seconds and the woodpecker flew off with exuberantly bounding flight, into the dark of the woods to begin collecting the next load. I walked on under the broad shady branches, pausing only to admire a Wren delivering his lung-busting machine-gun song from a branch on which a circle of sunlight fell, spotlighting the singing bird as he gave it his all.

CHAPTER 17

Aeshna isosceles The Norfolk Hawker

ʕ ʔ

Some of our dragonflies and damselflies are on the wing for months and months. You can be pretty sure that you'll see them any time in spring and summer as long as you're in the right habitat. That doesn't mean that individuals have particularly long lifespans, but that they have a 'staggered emergence', with new adults-to-be crawling out of the water regularly through the months. Others, though, have a more or less simultaneous emergence and so are available to the dragon-watcher for just a few short weeks.

These ephemeral species include the Norfolk Hawker. As well as having a short flight season, it has a very restricted range in Britain – you need to go to the Norfolk Broads to see it, and you must go between late May and mid-June. If this period happens to be unseasonably cold, windy and generally horrible, as it was in 2012, the pressure is really on. We booked a week's stay in Norfolk for the second week of June, and anxiously watched the weather forecast as the dates drew nearer but, weatherwise, summer seemed as far away as ever.

We drove up to Norfolk in squally rainshowers, and the mood in the car was sombre. Although I was almost sure it would be pointless I suggested we stop at Hickling Broad, on the north-east coast, on the way to our holiday cottage, and we drew into the Norfolk Wildlife Trust car park at about 3pm.

I had last been to Hickling back in 1995, to see my first Swallowtail butterflies. The best known of the special Broadland insects, the Swallowtail is a spectacular creature – huge, decked out in eyeball-popping yellow, black and blue, and it sweeps across its marshland habitat with imperious majesty and an amazing turn of speed. Hickling came up trumps back then, but when I looked at the sightings board, Norfolk Hawker did not figure on the list of Odonata. By then we'd paid the entry fee, so we went off anyway for a look around.

The layout of the reserve and its footpaths seemed quite different from 18 years ago. Time enough for a lot to change, I suppose. The broad itself is the largest piece of open water in Broadland, and is a National Nature Reserve, although like the other broads it came into being through the actions of people rather than nature. Peat and clay digging that began in Roman times produced a wide but shallow sheet of water, with a hint of sea salt, framed by the biggest reedbed in Britain. It is this reedbed that the reserve trails traverse, with a few birdwatching hides overlooking small pools.

A boardwalk led us out into the wide expanse of pale gold *Phragmites* stems, which were being thrashed noisily about by gusty winds. The occasional subdued Reed Warbler sounded a few squeaky notes from deep within — all sensible small creatures would be in deep cover on a day like this. The breeze picked up a female Azure Damselfly from wherever she'd been sheltering, and deposited her on one of the boards in front of us, where she sat clinging to the rough wood for dear life. Our Odonata list for the day started and finished with her.

We cut our walk short and carried on to our cottage. Sunshine was forecast for the following morning, and so we planned to rise early and go to a different reserve, one where we knew for certain that the Norfolk Hawker was present. The next day we were en route to RSPB Strumpshaw Fen before 10am. Another reserve that I knew of old, Strumpshaw is not dissimilar to the Hickling reserve in that it includes swathes of Broadland reedbed, but it also has wooded areas and flower-rich grassland.

We reached the visitor centre before any RSPB staff had arrived. Luckily there is a hide here overlooking a reed-fringed pool, so we whiled away the 10-minute wait with other early visitors, watching Swallows skimming the water and nesting Black-headed Gulls bickering among themselves. As promised, it was sunny and the breezes were light, though already wispy white clouds were beginning to coalesce. We'd have three hours at most to find our quarry before the weather closed in.

When the visitor centre opened we were first through the door, eager to find out exactly where to look for the hawkers. A smiling volunteer directed us to 'the meadow' and I raced off down the shady trail. Soon I was out of the woods and into a wide, grassy, flowery field, cut through with a small, poker-straight ditch 20cm deep, that turned through a couple of right angles as it traversed the meadowland. A profusion of sedges, waterweeds and other wetland plants grew on and in the water. Over this lusciously life-rich ditch, the volunteer told us, we would find our *isosceles*.

Rob caught me up as I walked very slowly alongside the ditch. Within moments a dragonfly appeared — brown, squattish, moving in a rapid stop-start flight just over the water. Our excitement quickly vaporised when it touched down on a stem and posed, revealing dark-marked wings and a short, tapering body. A Four-spotted Chaser, the first of many, it sat on its perch quite photogenically but was off as soon as a camera was raised towards it.

We soon got our eye in with the chasers and their particular shape and flight style. Then we began to notice Hairy Dragonflies as well, these keeping low, just above the water, and moving more slowly, less erratically. The blue males patrolled the ditch, while the yellowish females we saw were mainly egg-laying, squatting on floating vegetation with their abdomens curled down into the water. Seeing the two species side-by-side reminded me of a photograph a friend had sent me the week before, taken somewhere in Norfolk and showing a Hairy Dragonfly and a Four-spotted Chaser struggling together on a stony footpath. Apparently the Hairy had caught and eventually killed the chaser, in what must have been a real battle as the two are closely matched for size. No such drama seemed on the cards today, in fact the chasers were seeing off the Hairies when the two species met.

Watching the chasers got me wondering. Sometimes, as I walked along, a chaser would fly towards me from ahead or behind, and then slam on its brakes and hang midair alongside me. As it hovered perfectly on the spot, I could see every detail of the body that hung from the blur of wings. A green-brown abdomen, yellowish on the sides, tapered to a dark pointed tip: a square, powerhouse thorax. Most striking though were the eyes, dark and polished as those of a mouse or a bird. Unlike the pale, spotted eyes of most dragonflies, these ones looked like eyes I could relate to, and they looked like they were watching me. After a few seconds of this apparent scrutiny the chaser would resume its previous activity.

If you look at a dragonfly's head, you'll see eyes and a frons and not much else. There's no space in there for the physical framework of a towering intellect to be seated, is there? Yet these insects don't blunder cluelessly through their world. They have the wherewithal to catch agile flying prey, to defend territory, to find mates and, it seemed, to keep a careful eye on a big humanoid animal that may or may not pose a threat to them. Maybe the assumption that tiny brain = tiny brainpower is incorrect. Scientists modelling neural circuitry have found that systems small enough to fit in a honey-bee's head can be capable of complex processing and perhaps even consciousness. It is maybe not so far-fetched to suppose that these curious chasers were considering what, if anything, they should do about me.

Mostly, though, the chasers were interested in sex, and the prospect of sex instantly distracted them from whatever else they were doing. I saw numerous passionate encounters, as intense as they were brief. A courting pair would come together with a rattling clash of wings, and as they struggled to gain purchase on each other they'd tumble down to water level. Here I'd lose them among the emergent vegetation for a moment, though the sounds of their wings bashing against the plants and each other were still in evidence. Then the sounds would stop and the couple would re-appear, joined together in the familiar Odonata 'wheel', albeit a tight and uncomfortable-looking wheel, their squat bodies seeming unsuited to this degree of stretching and bending. The wheel position only lasted a few seconds, then the pair separated and seemed to go their separate ways, though I have read that males guard females while the latter lay their eggs. There were some females around doing just this, in flight with quick downward abdomen stabs at the water.

I walked slowly and Rob very slowly along the edge of the ditch. Being ahead, it was me who disturbed first a Common Frog and then a baby Grass Snake. Both dropped into the water and quickly swam out of view, much to Rob's dismay. Then, a crashing and rushing in the meadow beyond the ditch drew my attention to a far larger creature — I raised my camera to photograph the animal as it bounded away but as I only had a macro lens at the time, the photos were dreadful. Still, there was enough there that I could see the animal was what I thought it would be — a Chinese Water Deer, introduced to Britain in 1930 and for decades a Broadlands speciality, and the first one I'd ever seen. A small deer, only its head, neck and (when on the downward phase of its bound) chunky rump showed over the tall grass heads. It had a distinctive ginger tone and I knew that a front-on view would have revealed a broad, curiously piggy face with small beady eyes and, projecting downwards from the corners of its mouth, an impressive pair of straight, white tusks. But the deer was well and truly spooked and gave no backward glance as it raced for the shelter of the trees on the meadow's edge.

I walked back and forth along the full length of this ditch four times. Rob, being much more thorough, only did it once, but we were similarly unsuccessful in our attempt to find a Norfolk Hawker. All was quiet; few birds flew overhead. The clouds were joining forces and anxiety was building. Overheating in an unsuitable fleecy top I retreated into the shade of the woods for a sip of water and a breather. Here, I noticed yet another chaser by the path, this one poised elegantly at the tip of a long bare stem among a bramble clump, and went over for a closer look.

It quickly became apparent it was not another Four-spot this time, but something different. For a start, it lacked the 'four spots' — the prominent dark markings halfway along the leading edge of each wing, at the slight indentation there known as the nodus. Instead, it had a diffuse dark smudge on each wingtip. Its body was not the army-issue greenish-brown of a Four-spot, but a rather appealing rich honey-gold, marked with a dark central stripe. I edged closer, camera raised and clicking, and it showed no particular interest in me, making the occasional sally after a passing midge but always returning to the same favourite spot.

Although I thought I knew what it was, I had to check the book to be sure. To my joy I was right - it was a female Scarce Chaser, the 'other' species we'd hoped to find in Norfolk although they do also occur closer to home. With an unfortunate sexist bias I'd only properly learned how to recognise a male — blue, black-tipped body, dark triangle at base of hind wings, blue eyes — but this female took me aback with her beauty. I took photo after photo from various angles — sunlit, against the sun, from the side, from below — and the patient little dragon did no more than occasionally tilt her head (dark-eyed, like the Four-spots), a movement so crisp and sudden that it made me jump.

The Scarce Chaser male looks, at first glance, very like a male Black-tailed Skimmer. They are not in the same genus (the chaser is a *Libellula*, the skimmer an *Orthetrum*) but are a little fatter, with bluer eyes and the aforementioned dark marking at the hindwing base. However, females and immatures are quite different. While the skimmers are light lemon yellow, the chasers are rich orange, and the immatures are particularly dazzling. This has given rise to the species name *fulva*, meaning 'orange' or 'orange-brown' in Latin.

Happy with my find, I went off into the meadow again to find Rob, knowing he'd be keen to take some macro shots of such an obliging subject. He told me that he'd seen a Swallowtail but no notable dragonflies, and we walked back together, but on the very first stretch of ditch by the gate we glimpsed something over the water that stopped us in our tracks.

The dragonfly, for that was what it was, had cruised past and out of sight very fast, but I was left with an impression of something bigger and longer and browner than either a chaser or (the only other option) a female Hairy Dragonfly. We waited, and after a moment it came whirring back. Long indeed, and uniform light brown, smooth, balanced and powerful in flight, showing round, vivid green eyes like a pair of tiny Granny Smiths as it went past, it was a Norfolk Hawker at last.

Rob had a fast-focusing 300mm lens on his camera at this point, and I trusted him more than I trusted myself to catch a flight shot of this beauty. It seemed that a flight shot would be our only option, as the hawker showed no signs of needing to stop for a rest. After it had made a few passes along its preferred section of ditch we had worked out how to pick it out from the many chasers at a distance, and be ready to try for photos as it made its approach. One of the chasers, using as its lookout perch a white stick in the centre of the ditch, was not happy to see its larger cousin and kept chasing the hawker away just before it drew level with us, but we found a better spot to wait. Eventually the perfect moment came, as it slowed down and there was a rapid volley of camera clicks to my right.

The photos were good. Even so, we held out for a second perfect moment but in vain. Finally we swapped lenses and Rob went off armed with the macro to meet the Scarce Chaser, which I was sure would still be there on her stick (she was). I waited with the hawker and was rewarded when it actually settled on a reed. Unfortunately it was on the near bank and facing me so I could only manage photos of a bit of its underside. Not having an upperside view meant I didn't get to see the special fieldmark that gives the species its scientific name and relieves its body's otherwise rather boring brown colour scheme — the neat yellow isosceles triangle on its second abdominal segment. However, I did get close enough to properly capture a picture of those remarkable eyes.

I also noticed a number of blue damselflies deliberately flying at the resting hawker, behaviour highly reminiscent of the way smaller birds mob a predator. Of course, it is dangerous for a prey species to approach a predator, but if the predator seems not to be in hunting mode then the prey species will often try to chase it away, so that by the time it is ready to hunt again it has moved away to a different area. This was not the first time I've observed apparent 'mobbing' of dragons by damsels, though I was not completely convinced that the damsels really were trying to drive off the dragonfly, it seemed very sophisticated behaviour. But then I'd already noted the way the chasers seemed to watch me as I passed ... Maybe there is no way to be sure what was going on, but I knew that the more time I spent with both dragons and damsels the more they impressed me with their range of behaviour.

With our two target dragons seen and committed to memory card, we wondered what to do next. A week in Norfolk with no specific dragon duties (and no real possibility of seeing any new species) stretched ahead of us. We spent most of the week birdwatching along the north coast, seeing such delights as Spoonbills and Little Gulls at Cley Marshes, RSPB Titchwell and Sculthorpe Moor. The chance to watch birds from a comfortable hide was most welcome in a week of wind and rain and what the BBC weather folk described as 'disappointing temperatures for June'. A break in the weather halfway through the week did permit one more 'dragon day' and we tried a different part of the Broads to see if we could find Norfolk Hawker or Scarce Chaser (or both) again.

So we aimed for the RSPB's new reserve at Sutton Fen, and I refused to be daunted by the fact that the RSPB's own website stated that the reserve could not cope with 'large numbers of visitors' and was rather vague and coy about how to actually find the place. I navigated us to Sutton village, and from here we went down a series of increasingly narrow lanes, deciding on left or right turns more or less at random, and at last found a tiny car park by a loop of still water, while ahead our lane petered away to nothing.

It turned out that we had indeed found a reserve – the wrong one. This was not RSPB Sutton Fen but Butterfly Conservation's Catfield Fen, a site managed for Swallowtail butterflies but (presumably) also good for other Broadland wildlife. I was vaguely aware of Catfield Fen's existence but to stumble upon it now was a complete surprise. Still, it seemed silly to turn down this gift horse, so we began to prepare for a walk. The sign warned that to set foot on the fenland itself was to court catastrophe because of deep water, hidden holes and other hazards, but explained that a footpath called the Rond, which ran along the reserve edge on a raised bank, was a safe option for good views over the area.

While Rob unpacked his stuff, I checked the nearby water. Among the damselflies disporting themselves at the water's edge were a few Red-eyed, and while watching them it occurred to me that there was a chance I was looking at Small Red-eyeds, which do occur in the general area though usually emerge rather later than this. So I took some photos, looking for the tiny black cross at the abdomen tip that tells Small Red-eyed from its bigger cousin. It wasn't there – they were ordinary Red-eyed after all. I shrugged off this small disappointment and we went off to find the Rond.

The path led through a short stretch of woodland then out to the open fen. It was pleasantly broad and flowery, and there was a promising-looking ditch running right alongside it, with an impressive acreage of reedbed beyond. The weather was breezy, with fast-flying white clouds crossing the dome of the sky, switching the sunlight on and off and sending a kaleidoscope pattern of shadows across the far-reaching landscape. Not too breezy for dragons though – almost at once we saw a Four-spotted Chaser making its way along the reedy edge, and moments later the local celebrity — a Swallowtail, racing along out over the fen with very un-butterfly-like power and purpose.

The Rond path didn't give good views of the ditch; the reeds were too high in most places, so if there were low-flying dragonflies over the water here they eluded us. However, Four-spots were numerous along the path, and many of them were teneral, with striking glossy wings. One such individual detained Rob for a good 20 minutes — he knelt on the rather soggy grass and got busy with tripod and flashgun — while I kept walking, enjoying the extreme peacefulness of the place. A high, thin call from way above interrupted the quiet and I looked up to see a female Marsh Harrier circling above me. She called constantly as she wheeled in the blue and whiteness above, and I wondered if I was being mobbed by an anxious parent and if so what I should do about it. But the harrier soon drifted far away, and the quiet returned.

I found another teneral Four-spot airing its shiny new wings on a reed stem, and as it allowed me to approach closely I decided to see if it would get onto my hand. Figuring that to approach from below would seem less threatening than from above, I brought my hand slowly up under the reed stem, and once I was in position I touched the tips of its feet. The dragon immediately and willingly transferred itself from stem to fingers. I drew my hand back slowly, feeling clawed feet gripping strongly, and its almost imperceptible, perfectly balanced weight, and raised the dragon up to my eye-level. Close up, it seemed a miracle of engineering — half animal, half expertly assembled Airfix model. I only had a moment to take it in though, before it decided enough was enough and lifted off, flying strongly away on those perfectly fresh wings.

I remembered a recent conversation with a friend, who mentioned that her partner had a phobia of dragonflies being near him, ever since one either stung him when he was small. Now, it's easy to see why people might think a dragonfly *could* sting. The chasers in particular have wasp-ish tapering bodies that end in a sharp-looking point. The Broad-bodied Chaser, in its bright yellowy teneral form, is often mistaken for a hornet. The hawker dragons often have quite prominent appendages at their abdomen tips that could be mistaken for stings. The female Golden-ringed, with her long and stabby ovipositor, and her wasp-like black and yellow bands, looks like a stinging machine from hell to the uninitiated.

But looks deceive. No dragonfly can sting. They can, however, bite. The jaws that so efficiently chop up their prey may be used on a person in extreme circumstances, and while they don't inflict any pain to speak of their bites can certainly be felt. Odonata means 'toothed jaws', and a close look at a dragonfly tucking into some other poor insect reveals that these jaws, located below the frons, open sideways, with a panel at the top also moving up and out of the way. The whole lower part of the insect's face seems to open up, creating a surprisingly wide chasm. The fully open jaws show wicked serrations that aren't in evidence when closed.

I used to be dubious about whether a dragonfly really could bite a person, until I spoke to a bird ringer who'd had the experience of extracting the odd dragonfly from a mist-net set to trap small birds. Freeing anything from a mist-net requires great care and patience, as the mesh can entangle a small, struggling body very thoroughly. To free a dragon's wings and legs without damaging them would require controlling the insect's body carefully while picking the mesh away, an experience that no dragonfly would enjoy. In the course of doing this my ringer friend received a few dragon bites, which he said didn't hurt but felt like a slight pinch. While it isn't enough to do any damage, it would probably be enough to shock anyone not expecting it. This looks like another example of dragonfly brainpower — the awareness, somehow, that biting isn't just for eating but for hurting, and that attackers should be bitten if possible.

Walking on alone, I passed an open, muddy scrape, home to a pair of Lapwings that rose to scrutinise us from the air as we went by. Soon after this I reached the pretty, ruined brick mill, standing abandoned in the fen, which marked the end of the best part of the Rond. Here I waited for Rob and scanned over the reeds for Swallowtails, without success. I also explored the onward path but found it narrowed and became rutted, overgrown and the views became too obscured, so when Rob reached me I suggested we go back to the start. He agreed, but paused to take a few photos of the mill in its scenic and (periodically) sunlit surroundings.

Once again I took the lead, though stopped to investigate a striking yellow dragonfly that flew ahead of me and landed among some broken reeds near ground level. It was skittish — on my first attempt to get close I flushed it and it went on along the path a short distance. This pattern repeated itself a couple more times as I upped my stealth level, and eventually I got near enough for a good view through the lens. It was a Black-tailed Skimmer, first of the year. A light, bright yellow in tone, with a tapering abdomen like a slender cigar, it was marked with bold black 'biro' lines down its sides and across each segment join.

Like most of the other species in the chaser and skimmer group, the Black-tailed Skimmer undergoes a pretty radical colour change as it matures, or at least it does if it is a male. In females the colour just deepens and mellows, but males go from yellow (or green, or gold) to blue, developing a pruinescence or 'bloom' of dusty pigment, looking rather like that on a plum, which covers a body that's actually black. The male Black-tailed Skimmer has pruinescence on its first abdominal segments, fading away by segment 8 to leave a completely black 'tail'. However, the rigours of life in general and particularly of mating cause the bloom to wear away. Older male skimmers of all species often bear uneven 'mating scars', bare black patches up their sides where their bloom has worn away from being rubbed against their various partners. Females of these species, with no bloom to lose, stay beautiful for longer.

The Black-tailed Skimmer and its skinny little cousin the Keeled Skimmer are our only representatives of the genus *Orthetrum*. This name is a portmanteau of *orthos* — Greek for 'straight' and *trum* — Latin for 'tool', and refers to the straightness of the insect's claspers or anal appendages. Why this particular, rather obscure feature was deemed important enough to define the whole genus is anyone's guess — many other dragonflies have similarly straight 'tools', though others do have curved ones. The Black-tailed Skimmer's species name, *cancellatum*, seems still more baffling but this is one case where the Latin word doesn't mean what we think it does. In fact it means 'cross-barred' or 'latticed', which is a fair description of the pattern of black lines of the body of an immature Black-tailed, like the one I was watching now.

The rigours of mating were a few days away for this newly emerged skimmer, resplendent in fresh lemon and black. It was not yet a strong flier but already had strong ideas about approaching humans, and one careless move as I took its photo sent it careering away to find a quieter resting spot. It was replaced almost straight away by a male Hairy Dragonfly, which landed on the ground (very unusual for a hawker-type dragonfly) and allowed me to take my best-yet photos of this species.

The walk back produced a little more wildlife. A Peacock and a Small Tortoiseshell were both basking on bare ground along the path. Another Swallowtail flew up the path, with typical haste. Then in the reeds off to the left there was a sudden and tremendous panic-stricken crashing very close at hand. I watched carefully, as an obviously quite large animal went blundering away from the path edge, but could see nothing of it whatsoever. So while I strongly suspect it was a Chinese Water Deer, I couldn't be certain.

Finally we were back at the start of the Rond. This last (or first) bit of path veers away from the fen and goes through a shady glade, but there were sunny clearings here and there, and among the vegetation were numerous damselflies. We stopped for a closer look.

Many Odonata species actually head *away* from water when they are freshly emerged, and spend their first few days of adult life in places like this. In some species, the females spend practically their whole lives a good way from the nearest stream, lake or pond, and only come to water when they are ready to mate. (Males tend to return to watery habitats sooner, so as to be ready for when any female turns up.) The damsels frequenting this little woodland patch were mainly adult females and immatures of both sexes.

One of the few adult males present caught my eye. A blue damsel, he seemed to have more extensive black marks on the abdomen than a Common Blue or Azure. A closer look revealed broken antehumeral stripes and a complex wineglass mark on the second abdominal segment ... but even without these very specific field marks the generally dark impression was noticeable. He was a Variable Damselfly, the first of the year and in fact only the second of my life, following the one I had found in 2011 when I was supposed to be birdwatching.

Although Variable Damselflies are indeed variable, these traits of the male are consistent enough that identification isn't that difficult with a clear view. I wasn't so sure about the females, though I knew that they came in various different colour morphs, so needed to do a little quick revision, as we'd soon established from checking the males that there were Azures present as well.

Azures and Variables are the two commonest British species in the genus *Coenagrion*, and are very similar in every way. Separating them in their various forms was an interesting exercise (for me, at least. Rob just wanted me to point out which ones he should photograph). It turned out that that the males were trickier to identify than the females. Female Variables don't come in a green form, which meant any green ones had to be Azures, while Azures don't come in a dark form, so the mostly black ones were female Variables. The only overlap came with the small number of 'blue phase' females, their abdomens marked with equal width bands of blue and black along the whole length, like an Inter Milan football scarf. Such females occur in both species, but the blue phase Variables have green markings on their thorax, while the Azures are marked with blue.

I carefully picked up a drab-coloured teneral damsel for a closer look. It was very fresh and showed no particular alarm at being caught — it even posed for photos on my hand, walking ponderously from finger to finger with its long abdomen trailing over my skin. With close scrutiny I could make out the typical black pattern of a male Variable, but the parts that are bright blue in a mature male were a very dark violet. After its explorations, the damsel sat completely still, legs bowed so his body was in contact with my hand, perhaps picking up some warmth from me. Then the fine, shining wings suddenly blurred into motion and he lifted off, to be carried away into the vegetation on the breeze. Both cloud and wind had picked up — it was time to go.

CHAPTER 18

Orthetrum coerulescens
The Keeled Skimmer

ꙨꙨ ꙨꙨ

It was only late June but the relentless cold and rain gave every day an autumnal feel, as though summer was coming to an end before it had even begun. Plan after plan was cancelled, weekends written off, our chances of finding the few remaining British dragons and damsels that we had yet to see were dwindling away.

That wasn't the only thing that was dwindling away. Dragon-hunting aside, things had not been right between us for a long time, and as June came to an end Rob called time on our relationship. After nearly seven years together, this was a tremendous wrench, but there was no anger in our final talks, just sadness, and we hoped that we would be able to keep our strong friendship alive somehow. To that end, we decided that two more dragonfly trips would take place as planned.

So the next day I went to live temporarily with my friend Sue, sharing her pretty cottage in a tiny village near Maidstone, and Rob stayed at his flat in Sevenoaks. I would be returning to Sevenoaks in due course, but my own flat was let to a tenant at the time, who needed two months' notice before I could reclaim it. So I had to find another place to go in the meantime.

A week or so after the split we had an awkward meet-up at Sevenoaks train station. From here we drove west, taking the M25 and then M3 towards the New Forest, in search of Southern and Scarce Blue-tailed Damselflies. We talked for some of the way, about nothing important, and at times it was almost as though nothing had changed. At other times, there was quiet between us and the quietness had a particular density and quality that said, actually, yes, everything has changed. Except for the dragonflies and damselflies, that were (hopefully) still waiting out there somewhere to be discovered.

The weather had taken pity on us and it was a pleasant enough day, not too windy, with the promise of some sunny spells. Certainly more promising than when we had made this same trip a couple of weeks ago, when there'd been cloud and a cold breeze — and we compounded our problems by following the path the wrong way and ending up in habitat wholly unsuitable for either of the damselflies we'd hoped to see. Negotiating the complexity of motorways around Southampton, we ended up on the A31, driving through the starkly beautiful open heathlands of the New Forest. We turned off down a minor road to find our way to Rhinefield Ornamental Drive.

On our first visit, Rob was outraged that so much of the New Forest should be so un-forested, but we were both enchanted by the beautiful ponies that wandered at will on both the heathlands and around the towns and villages of the forest. We were less enchanted when a large group of them, a motley bunch in assorted colours and sizes, formed a roadblock across the narrow lane and stood there peacefully, regarding us with melting dark eyes and gently rotating jaws as they processed their last mouthful of fodder. We waited a little while, as other cars began to draw up behind us. Then Rob said, 'They're scared of people, right?', got out of the car and walked towards the ponies, a move which didn't impress them in the least. A brief standoff ensued, Rob having approached as closely as he was willing to, and the ponies not yielding as much as a millimetre. Then he returned to the car, got in and wordlessly turned it around. We found another way.

The car park called Puttles Bridge was our destination, on the edge of a part of the forest that is

actually quite replete with trees. A stream cut through the woodland and the heavy rain had created extra pools and puddles along its banks. On the opposite side of the road, the terrain opened right up and the stream weaved through a heathery valley. This was — we were almost sure — the Silver Stream, our destination. But before I reached the road I was distracted by the sight of a small yellowish dragonfly flying low over a large clump of bilberry, and I stopped for a closer look.

This patch of low, scrubby greenery turned out to be alive with Odonata, of two species. The yellow dragonfly was a Keeled Skimmer, an immature, and there were dozens more exploring the same small area, including one or two more mature males with powder-blue pruinescence appearing on their slim, tapered abdomens. The other species present was another acid-heath specialist, the Small Red Damselfly. Again, there were lots of these, mostly dark immatures but the odd adult male with its delicate body pure red from base to tip.

I knew both species from Thursley Common, much closer to home. I'd always found Keeled Skimmers easy to see there, but had only finally caught up with my first Small Reds last summer. The two species are both restricted to this particular kind of habitat, using the dark peaty pools, streams and channels that lurk among the heather and rough grassland to lay their eggs. In their early adult life, though, they avoid the wettest areas and head for drier ground. This small clump of bilberry was a veritable Swiss finishing school for these adolescent Odonata, and was teeming with food for them in the form of other, smaller, flying insects.

Rob joined me and we stalked around the outside of the bilberry clump, looking for good photographic angles on the damsels and dragons as they flitted about. They were uncooperative, choosing obscured perches or awkward angles. I gave up after a while and instead pointed my lens at a gloriously golden, very fresh male Large Skipper butterfly, which had picked a nicely prominent high leaf as his territorial perch. From here he was ready to launch himself, with anger or ardour as appropriate, at any other passing Large Skipper, but for now he was the only one of his kind in the immediate area so he just posed and waited.

We finally moved on, crossing the road to inspect where the stream emerged from Puttles Bridge and made its fragmented way across the heath. Finding a way down to the water's edge was a matter of careful stepping and hopeful leaping to get over the numerous puddles and boggy patches, but we got there in the end, and made our way along the stream as it came through a clump of Scots pine and then broke into separate strands as it crossed the heather. All the way, we inspected the open water for insect activity, but there was none.

So, where were these damselflies? After our original failed trip, I had checked and rechecked the whereabouts of 'Silver Stream', home to both of the species we hoped to see, and felt as sure as I could be that we were in just the right place. The weather seemed good enough for them to be flying (and the activity of the Keeled Skimmers and Small Reds we'd already seen would seem to bear that out). Both species had been seen within the last week. But for us, there was nothing.

Rob thought we might be in the wrong spot, and in the absence of damselflies we jumped and sidled and clambered up the slope to get a look across a wider stretch of heath. From our vantage point on the ridge east of the stream, we could see plenty more of the New Forest, the dark undulations of heather stretching away on all sides, and the bright, sparkling and silvery waters of what may or may not have been Silver Stream marking out the lowest point of the valley before us. There were no other streams, silver or otherwise, in sight. The heath was immensely quiet, apart from the faint and mournful calls of an unfeasibly high-flying Curlew away to the south-east, just a speck as it circled and whistled against a bank of smoky-grey cloud.

We had a bit of a conference, and decided to go up the road a little further and see what we could find. The terrain became more wooded as we walked along the roadside, and although we cut into the woods where there was a trail, none of the trails led to anywhere likely-looking. So we turned back, but straight away noticed something very strange, that occupied us for the next half an hour.

Dangling from one of the trees near the road was what looked at first glance like a gigantic clot of brown and blue-white chewing-gum, looped over a branch by a long, drawn-out strand and swaying gently in the breeze. It was only the size of it that compelled me to look more closely, as surely there was too much of it to fit in a normal human mouth.

As I got closer I saw that it wasn't chewing-gum at all but a pair of massive Leopard Slugs, dangling from a thick thread of slimy mucous and twined intimately together in what some books (including this one, it seems) coyly and euphemistically refer to as a 'nuptial embrace'. There was nothing coy about what was going on though — both slugs had their most private parts fully exposed and meshed together. As hermaphrodites, each slug has the same equipment, usually described as a 'penis' (presumably just because it goes out rather than in). I can report that the penis of a Leopard Slug is oddly beautiful — pale blueish, somewhat translucent and frilly-edged, rather like a delicate jellyfish. It is also very big relative to the slug itself. But no penetration occurs — the slugs exchange sperm through direct contact between their sticky-out bits.

We watched and photographed this spectacle for an indecently long time, feeling an unsettling mixture of fascination, embarrassment and revulsion. The slugs were oblivious, slowly twirling together and dripping goo. Then they both quite quickly winched in their respective apparatus, and when that was complete very slowly disentangled themselves and began to climb up their slimy rope ladder.

Why does the Leopard Slug do its thing while dangling in midair? Not all slugs do. I suppose that they are safer up there, a good metre above the ground, as most slug-munching predators will be looking for them in the undergrowth. Leopard Slugs are sizeable molluscs: just one would make a good meal for a Blackbird or Hedgehog, so two would be a feast.

With each slug now fertilised by the other, they would now have to survive another four days or so before seeding the next generation by laying their eggs (more than 100 of them, apparently).

After the slug experience it was impossible to write off the day as a disaster, but we had still not found our damselflies. Paranoid that we'd somehow got our directions wrong again we drove into Lyndhurst, the nearest town to do an online check. We also perused the Brooks and Lewington book. Both sources seemed to confirm that we had gone to the right place. It was the damselflies that were in the wrong place. On the book's advice, we decided to try to find the other end of Silver Stream, off the A35.

We found the stream and a place to park nearby, alongside a gate complete with stile. The foot of the stile on the other side was under around 20cm depth of water, and the whole immediate area was flooded. I knew to my cost that my inexpensive new walking boots had all the waterproof qualities of a teabag, and even if they hadn't the water would have come over the tops. I sat on top of the stile and wondered what to do, while Rob, carefree in more expensive, extensive and durable footwear, patiently waited. Finally I removed both boots and socks, rolled up my trousers to my knees, and stepped into the surprisingly chilly water. I retrieved my footwear from the stile step and began to wade streamwards — there were at least 10 metres of flooded ground to cross before things dried out a bit. Underfoot, saturated grass mixed with soft, oozy mud worked its way more thoroughly between my toes with every step, but the main feeling I had was one of liberation.

The stream rushed fast and clear across the field and under the bridge across the busy road. On our side of the road its path was lined with small trees and its banks were a little bare. Rob climbed up to the road and took a look on the other side, and came back to tell me that it looked 'really good'. I climbed up onto the bridge, dried off my feet as best I could, and put my footwear back on to cross the road. I made a dash across the road and, from the other side, saw that Rob was right about the stream. It twinked off into the distance across an open, lush field, and its margins had good amounts of vegetation here and there. The ground on either side even looked reasonably dry.

We picked the west side of the stream, climbed down the other side of the bridge and began to walk, slowly and stealthily, alongside the stream. We'd been walking for just a few minutes when Rob spotted a damselfly, but it was the wrong colour – a 'red' rather than a 'blue'. In fact it was a Small Red, the same species we'd met earlier in the day. An adult male on the search for a mate, it flitted along the shore, showing off a slim length of pure red body. Noticeably smaller and redder than the much commoner Large Red Damselfly, it was a beautiful little insect, but it wasn't a Southern Damselfly.

We saw plenty of Small Reds in favoured spots along the stream, but it was a while before we saw our first 'blue'. Then suddenly there it was, unhurriedly bobbing along the stream edge, oblivious to our squawks of excitement. We tracked it for a surprisingly long time as it lazily danced around, bobbing showily up and down, but finally it touched down on a sedge close to the fast-flowing water, and I raised the camera and began to take photos, and to study our quarry as I did so. It was pale turquoise, biggish, and its thorax sides were beautifully marked with two fine black lines and two fine blue ones. I put my camera down with a sigh and broke the news – 'It's a White-legged'.

With scant regard for our disappointment, the male White-legged Damselfly posed prettily on his perch. Although not the species we'd hoped to find, he was well worth a second and third look, especially as last year we had seen only one of his species. The White-legged Damselfly is so named for its broadly flattened and strikingly white tibias – the middle joint of the leg before the terminal claw. All six legs sport these peculiar tibias, and the male damsel shows them off by letting them dangle down when he is in flight. The reason for these elaborate legs and the way the males show them off is sex – the dangling legs and bobbing flight constitute a courtship display, and, according to the field guide, the fun doesn't end there. Once the pair have come together, the males' enthusiastic groping of their partners with the expanded tibias provides the 'tactile stimulation' that is necessary for effective copulation.

Another trait of this curious damselfly is the exceptional width of its head. It really is the hammerhead shark of the damselfly world, the space between its bulging eyes easily exceeding the width of each individual eye. I recalled a summer evening years ago, when I'd joined my then-boyfriend on a fishing trip to some quiet pools somewhere near Tunbridge Wells. In the long grass by the water, damselflies were bobbing about and settling, among them several White-leggeds (not that I knew the species at the time). As I leaned in for a closer look at one of them, it clearly noticed me and didn't care for the way I was looking at it, as it shimmied around its grass stem perch so it was on the opposite side, a nifty trick I've since observed in other damsels. This move meant its body was completely hidden, but the entire sphere of each eye was still visible, peering at me comically from either side of the grass stem. Despite its suspicions it was sleepy enough to allow me to carefully pick it up and I studied it closely — weird legs, weird eyes and all. It was easy to identify it from my old *Reader's Digest* insect book at home later. I'd assumed at the time that the species must be very common, but after that encounter it was many years before I saw another.

Last year's White-legged Damselfly had been a female, seen on the Thames during a fruitless search for the Common Clubtail in June. Clinging to a lily bud just clear of the water in a small inlet with her abdomen tip submerged in the water, she seemed to be egg-laying but there was no male in sight and this species, like most damsels, usually lays with her partner still attached, in the tandem position. Perhaps she was just having a wash. She was small consolation on that disappointing day, but I later found out that, on the Thames and some of the other large rivers that it frequents, the White-legged is in trouble. Increased boating traffic, especially bigger and faster crafts, creates lots of backwash to the shore, and this is taking its toll on the bankside vegetation. But White-legged Damselflies need emergent plants to climb up when they begin their adult lives.

Here on this New Forest stream, the only aquatic traffic around was the occasional gang of ponies having a paddle, so the White-legs should prosper for years to come. But of their scarcer cousins, the Southern Damselflies, there was still no sign.

The Southern Damselfly belongs to the same genus as most of Britain's blue damsels — *Coenagrion*. It is one of only two British Odonata species (the other is the Norfolk Hawker) that have protection under the Wildlife and Countryside act, meaning that if we'd been lucky enough to find one, we wouldn't have been allowed to take it home with us — not that we wanted anything more than a sighting and maybe a photograph. This protected status reflects the species' very limited distribution ... but from accounts I've read, on some New Forest streams at the right time of year, the Southern Damselfly outnumbers all other species. It is a small *Coenagrion*, smaller than its common relatives the Azure and Variable Damselflies anyway, and the males bear a lovely 'Mercury' mark on the second abdominal segment. In case you don't know what a Mercury mark looks like (I didn't), it's like a stalked X, and denotes the Roman god Mercury. The species name *mercuriale* refers to this marking as well, though it works equally well as a reference to the damsel having a mercurial, unpredictable nature.

We walked along the stream as far as we could go, before floodwater took over the ground. Rob was keen to try the other side of the stream, where our cast shadows would be thrown away from the water rather than across it and be less likely to spook any damselflies. So off came my shoes again and I waded across the stream barefoot, wincing and wobbling as I stepped on the sharp gravelly stones. Once across, the problems had really only just begun, as I now had to find a way through the very boggy grassland here. There were holes much deeper than I could roll up my trouser legs and I sank thigh-deep into several of them. Finally I got close enough to Rob (who'd found a much less hazardous way across) to pass him my camera, and free of this expensive encumbrance I enjoyed the rest of my wade through the bog onto higher and more solid ground. Tired and sore of foot, I found a spot to sit on the bank and let my bare feet dangle in the cold stream water, and a Small Red Damselfly took the opportunity to settle on my right knee and bask in the thin sunshine.

The shadows were slowly growing, and I felt sure that we would not see a Southern Damselfly today. The same went for the other damselfly that we had hoped to find today, the Scarce Blue-tailed. This little damsel rejoices in the scientific name *Ischnura pumilio*, translating as 'thin-tailed dwarf'. But it didn't look as though we were going to be in a position to judge whether it deserved a better name than this.

My field guide advised me that the Scarce Blue-tailed Damselfly has 'very specific habitat requirements', liking small, shallow pools with little emergent vegetation, and many suitable patches are rather short-lived, being things like earth extraction works and even puddles formed in vehicle tracks. If the pool dries up, or its margins become colonised by vegetation, the damselflies don't like it any more. A colony can therefore be there one year and gone the next, and opportunities for the species to spread and colonise new places may be thin on the ground. This is in stark contrast to the ordinary Blue-tailed Damselfly, which cheerfully lives and thrives anywhere and everywhere they have access to established still or slow-flowing water.

Because of this uncertainty, I had from the start been more concerned about seeing Scarce Blue-tails than Southern Damselflies. I'd even had a few anxiety-filled dreams about them. In my dreams it was not the adult Scarce Blue-tails that I saw, but immature females of the colour form *aurantiaca*. Adult Scarce Blue-tails are very similar to adult Blue-tails: you need to minutely examine the extent of blue on the tail to make an identification. But the young *aurantiaca* female is something else – a dazzling and unmistakeable creature with a bright fiery orange thorax and body sides. She is one of the prettiest of all damselflies if the illustration of her in the book is to be believed, and the book said we would find her here, in the same areas as Southern Damselfly. But in all our wanderings we hadn't even found any habitat that looked suitable, let alone any sign of the damsel itself.

These two elusive creatures would have to wait for another day and another year. Rob was on call overnight and so needed to be near his workplace by 6pm. We decided to head back, but to stop at another guide book-recommended stream on the way.

As soon as we pulled up by this last-chance stream, I knew that today really wasn't going to work out well for us. Although surrounded by open heath, the stream itself ran through a corridor of trees – its water would be shaded and thus unattractive to damselflies. Nevertheless, we walked alongside it for a short way, along narrow twisting pathways through tough, steely heather as high as my chest. Just down the hillside ran the stream, in shadow. We came to a halt and in the quiet the vague, hovering sadness of the day became more strongly felt. I looked up into a now clear blue sky, and watched a Buzzard float overhead on a wide spread of patterned wings, some unidentifiable bit of half-dismembered prey dangling from its talons. With all useful words exhausted, we made our way quietly back to the car and began the long drive to our homes.

CHAPTER 19

Sympetrum sanguineum
The Ruddy Darter

≈❧ ❧≈

My weeks living with Susan in East Sutton passed quickly. I knew I was, in theory, missing out on valuable dragonfly days but the weather was more or less uniformly horrible through the end of June and into July. I would draw my curtains at dawn and there before me were the lovely rippling green valleys of the Greensand ridge, being rained on.

I'd dress, pull on trainers and go off for a run down the lanes, because in my opinion running is the only enjoyable thing to do outside when it's raining, and often I'd see lots of wildlife. A Badger, crossing the road with its lurching, shambling trot. The elongated ginger shape of a Weasel running hard, rippling along the edge of the lane and disappearing among the bases of the towering nettles that lined the verge. A noisy family of fledgling Kestrels testing out their wings around the tower of the village church. The round lump of a Little Owl, crouched and scowling on the tile-hung roof of a beautiful old farmhouse. Piping Bullfinches, trilling Yellowhammers, tinkling Linnets. Proper countryside wildlife was everywhere, and its presence did help soothe my spirit.

The cottage garden was another source of enjoyment. Sometimes, when there was a rare half-hour of sunshine, I'd stop working, hurry out there with the camera and take pictures of the flowers. As July progressed, the weather gradually picked up and butterflies began to appear in the garden — Meadow Browns, primarily, broad-winged, dark and velvety, slurping the nectar out of bramble flowers. There were also Small Skippers, tiny and golden, buzzing like nimble little moths through the maze of grass stems. It was also proving to be a good Red Admiral year, and I saw more and more of these big, imperious butterflies gliding through the garden, in search of buddleias and early windfall fruit.

This rural idyll was all very well but I did hanker to go further afield, and to see some Odonata into the bargain. So I emailed Phil, my blogger friend, who lived fairly near, and suggested a trip to Dungeness. We chose a late July day in the midst of a spell of proper, hot summer weather. Thanks to our incessant talking we missed a few turnings and it was about 10am when we finally scrunched into the car park at the main RSPB reserve, by which time a heat-haze already shimmered over the expanse of vegetated shingle.

There were only a couple of other cars parked — Dungeness is a birder's nature reserve, but July is no month for birdwatching. In high summer, birds have mostly finished breeding and have begun their annual moult. Many species become almost invisible, so low a profile do they keep in the weeks of less than optimal feather condition. With spring migration over, autumn migration yet to begin, and little chance of weather-related wanderings, there is little chance of finding any rarities. No wonder so many birdwatchers turn their attention to other groups of animals at this time of year. Many a dragonfly fanatic started out as a bored birdwatcher, looking for something to watch on a quiet day in July.

We began our walk alongside the shingle ridge that screens off the large Burrowes Pit from the path. Various hardy little plants grew out of this bank, pushing their way between the pebbles with impressive tenacity. We noticed a tiny butterfly flitting around one such clump of plants and crouched down to examine it when it settled. It was a Brown Argus, a pretty creature with dark chocolate uppersides lined with little orange half-moons, and an intricate pattern of white-circled spots and more orange half-moons on its light fawn underside. It rested with wings closed shut — butterflies use their wings to maintain the right body temperatures, and on hot days like this are not going to sit around soaking up extra warmth.

The pit is a long, deep excavation into the shingle, and as you walk along by it you pass a succession of birdwatching hides. Come here in winter, armed with a telescope, and you stand a good chance of finding all kinds of goodies out on the deep water — rare grebes, divers and ducks are all possibilities. In summer, not so much. We did have a quick look but found very little — a Ringed Plover, a few clusters of Tufted Ducks, a glossy Carrion Crow stalking across the down-slope of the shingle bank.

Reaching the far end of Burrowes Pit, we turned right and from the landward side came a big flock of Lapwings, looking to be on a course directly overhead. We stopped to watch them as they came over. Lapwings are beautiful birds, smartly dressed in black and white that shows colourful iridescence in good light, but these were looking far from their best, with many of their flight feathers obviously worn and some missing altogether. Behind them came another, larger flock, but these were Starlings, not Lapwings. They landed on the shingle bank ahead of us and began strutting around and stabbing at the ground with their stout and pointy bills. Most were brown juvenile birds, in pretty good nick featherwise and certainly in good voice.

The general shortage of birds was no surprise but there didn't seem to be many insects on the wing either. I did stop to check a skipper butterfly that was feeding on the lovely blue flowers of viper's bugloss, and identified it as an Essex Skipper. This little butterfly is very similar to the Small Skippers I had seen in Susan's garden. The best way to tell the two apart is to look very carefully at the undersides of the antennae tips — these are brown in Small and black in Essex. Also, bucking the Essex stereotype that seems to prevail these days, the Essex Skipper is a little *less* orange in tone than the Small, but a slightly subtler straw-like colour.

At Denge Marsh we turned left and followed the river in a vaguely westward direction. I soon twigged that this was the very same river where

Rob and I had seen the Vagrant Emperor dragonfly back in April 2011, and soon we would reach the little footbridge. But today there were no dragonflies to be seen, just a few Six-spotted Burnet moths visiting riverside flowers. More juvenile Starlings swirled around, and up ahead a Corn Bunting sat stolidly on a fence post, eyeing us as we approached. I stopped to take its photo — it was a handsome-looking bird, thickset and chunky of bill, its dark eyes very prominent in its sandy-coloured face.

We reached the footbridge, and paused for a look around. Over at Denge Marsh, a lone Marsh Harrier glided low over the water, a dark, graceful shape on slightly uplifted wings, with a striking white crown. Down below, a Pike loitered in the river, just as had happened last time I was here. But that had been a big bruiser of a fish and this one was a mini Pike, though its wedge-shaped head looked proportionately very big. Phil, a keen fisherman, explained to me that little Pikes like this are called 'Jack Pikes'. Small though it was, this Jack looked every inch the fierce and fearsome predator with its lantern jaw and large, coldly gleaming eyes. We watched as it swam slowly downstream, losing itself among the thick underwater vegetation that lined the shallows of the little river.

We headed back, beginning to really feel the heat now as we tramped along the level, grassy path. A Carrion Crow flew over, bill open wide as it panted out some excess body heat. When we reached the Denge Marsh hide we were both more than ready for a break, and the interior of the hide was blissfully cool.

Opposite, well out in the lake, was a large floating plastic structure, oblong in shape and topped with a wooden frame filled with shingle. It was a tern raft, provided as a safe nesting place for Common Terns, and it was being used for exactly that purpose. A dozen or so adult Common Terns were standing about on the raft, having divided up the available space so each pair had a small but fiercely defended nesting territory. Among the sleek, white adults pottered a few young chicks: grey, fluffy, rotund and softly dappled. There were also adult terns out hunting over the water, describing graceful arcs as they rose and dipped, wheeled and turned. When one returned to the raft, all of the perched terns would stop what they were doing to screech a greeting, and the incoming bird would bow and posture to its mate.

We watched the terns for a good while. Then it was back out into the heat of the early afternoon. On the last stretch of footpath, at last, we began to see dragonflies. There were suddenly darters everywhere, around and alongside us, catching our attention with their rapid dash-pause-dash flight. They stopped often to stop on the ground or on vegetation, allowing me to take some photos.

The first darter I photographed was a male Ruddy Darter. He was a real beauty, his slim and shapely spindle-shaped abdomen an intense, deeply saturated crimson red, his spread wings swept forward like a sun-shading parasol, his great eyes chestnut and expressive-looking with their large pseudo-pupils, his spiky little legs jet black. He was perched on a head of viper's bugloss, its intensely blue flowers a clashing contrast to his own vibrant redness.

The Ruddy Darter is a common species in south-east England, becoming progressively scarcer as you go north and east, but like so many other southern species it does seem to be spreading in the UK. Its species name *sanguineum* comes from the Latin word for blood, and the intensity of redness is one way to tell the Ruddy from the Common Darter, though the body shape and all-black legs are two better ways to do it. Male darters tend to become redder with age, so an older male Common Darter could look just as blood-hued as a younger male Ruddy.

This last part of the path, leading through more vegetated, less shingly areas with stands of gorse and patches of grass, proved to be Dragonfly City. Besides the many darters we saw a few hawker dragonflies, big, long, fast beasts that flashed by too quickly for even a wild stab at identification to be possible. The darters were mostly Ruddies but there were a few Commons around as well.

Male Common Darters are longer and straighter than male Ruddies. Their less waisted abdomens are a lighter, more orange colour, and their faces are also less flushed with red. But females of the two species are more similar, in both colour and shape — both are yellowish, darkening with age. It really does come down to examining the legs, which bear a yellow stripe in Commons.

The Common Darter lives nearly everywhere in Britain, and is quite a wanderer. It will also use very small ponds for laying eggs — I've even seen a tandem pair laying in a tyre-rut puddle. Its abundance, plus its habit of warming up by sitting on anything made of wood, makes it one of our most familiar dragonflies. It's also a species that you'll see long after most others have disappeared — as long as there are enough mild autumn days for it to keep feeding it may survive into November. The species name, *striolatum*, means 'fine lines' in Latin, and may refer to the narrow dark divisions between its abdominal segments, but I'm guessing here.

We explored other parts of the reserve for the rest of the day. At the ARC pit, another flooded gravel extraction site we found numerous darters and got a better look at some hawker dragonflies — a couple of Brown Hawkers, wings vividly gold in the sunlight, and an early male Migrant Hawker, a sturdy, mid-sized, mostly blue creature that was stop-starting his way along the shore of the lake, watching out for a female to pounce on. Out on the water, among the motley crew of moulting ducks bobbed a pristine juvenile Garganey, small, elegant and dainty, looking like a cat-walk model picking her way through a crowd of scruffy builders.

Ending a lovely day out is always tough, but much more so if it's the first day out you've had for weeks and the next one won't be any time soon. We dawdled back to the car, stopping to admire a mixed flock of gulls loafing on the shingle until the gulls got nervous of the way we were watching them and started to shuffle away. Over the smaller and more vegetated pool to our right, a male Emperor Dragonfly swept to and fro, a spectacular, flashy streak of green and blue, making the most of a glut of midges that the hot weather had summoned out of the depths of the lakes. While 2012 had been a tremendously disappointing year up until now with so much rain, cold and gloom, it was starting to look as though a little bit of summer had been held back on reserve.

CHAPTER 20

Erythromma viridulum
The Small Red-eyed Damselfly

⧽⧼ ⧽⧼

By mid-August, I was living in Sevenoaks again, in my own flat, and was kept busy putting up shelves, changing locks and settling my cat into her new home. I was keeping close tabs on dragonfly and damselfly news but time, money and travel difficulties meant that I couldn't travel too far from home.

I am an avid reader of the blog posts from Howard Vaughan, reserve manager at the RSPB's Rainham Marshes reserve. This little gem of a place, on the banks of the Thames in an otherwise unprepossessing part of east London, has a fantastic track record for attracting wildlife of all kinds, and Howard's regular updates on the RSPB's community pages bring this good news into the living rooms of those who can't get out as much as they'd like. One of these blog posts mentioned that both Red-eyed and Small Red-eyed Damselflies were currently in evidence on the reserve, and as I had yet to see the latter I hatched a plan to visit.

Two partners-in-crime agreed to come with me. Shane, all smiles and camo gear, picked me up in Sevenoaks, while pulling into the car park at Rainham, we spotted Graham, a rangy bearded figure in trademark baseball cap, waiting by the steps up to the visitor centre.

Shane, Graham and I have met on several occasions now, after getting to know each other via the RSPB community forums online. I am usually horribly nervous about meeting anyone new but wildlife people aren't like normal people ... the shared obsession we have with watching and taking pictures of anything that moves makes conversation instantly easy and natural. All of us were toting cameras — Shane a practical and compact 'bridge' model, Graham a Canon DSLR with 100—400mm zoom lens, and me with the usual Nikon plus 300mm prime lens — the heaviest, least practical set-up of all. In my camera bag was the 180mm macro, just in case a Small Red-eyed Damselfly chose to pose at close range for me.

It didn't look as though we'd stand any chance of even seeing a damselfly, though. Sludge-coloured cloud hung low over the marsh, spitting squalls of rain. But things were forecast to improve. Looking out from the glass frontage of the visitor centre, you can see the whole sweep of the reserve, a scrap of beautiful wild marshland flanked by urban sprawl, the new A13 flyover, a towering rubbish tip and, to the south, the tidal Thames. Only Rainham's history as an MoD firing range saved it from the intensive development that has transformed most of the land alongside the outer reaches of the river. Back then it was rough ground with a few pools, which attracted many common and a few rare birds for the handful of intrepid birdwatchers who squeezed through gaps in the fence to go looking for them. The RSPB bought the site in 2000, but it took years of work before the reserve could be opened to visitors.

Fence-squeezing was not necessary in 2005, when a Sociable Plover turned up at the site. Even though this was before the RSPB had officially opened the reserve, access was arranged for birdwatchers to see this eye-wateringly rare bird, native to Asia and with a Critically Endangered global status. In years to come, more rarities would arrive, including Britain's first ever Slaty-backed Gull in 2011.

For all its promise, Rainham Marshes was no easy prospect for the RSPB. The amount of hard work that has gone into transforming the landscape — creating shallow pools, channels, reedbeds and other watery goodness — has been immense, as has the creation of birdwatching hides, boardwalks, viewing platforms, school study areas and wildlife gardens.

Then there were the human problems. A small and disreputable element of the local populace objected to the RSPB's plans, and made life difficult for workers on the site for a while. Even now, with the reserve completely encircled with high fences, there is still the occasional episode of nocturnal vandalism — a few months before, persons unknown had entered the reserve after hours and damaged the lining of a wildlife pond so badly that the pond had to be drained and its occupants — assorted leeches, water beetles and damselfly nymphs — relocated to safe new homes.

The only humans on view today were a typical cross-section of bird-watching types. After a quick look at the sightings board — an impressively colourful map of the reserve on which interesting animals were pinpoint-ed with magnetic representations — we set off too, down a long sloping bridge to the start of the circular trail.

Alongside the path grew a profusion of colourful flowers, in particu-lar the beautiful blue chicory. Tall, with dry-looking, spiky dark stems bearing little in the way of leaves, the chicory plants were liberally cov-ered in big, light blue, daisy-like flowers with long, fringe-tipped petals. I lingered to take photos, even though I knew that my camera would not catch the delicate colour quite correctly. On the opposite side of the path, Shane and Graham studied a bank of bramble, among which one or two chocolate-and-orange Gatekeeper butterflies lazily flitted. Despite cloud and drizzle, the air temperature was just high enough to allow them to fly.

Following the trail anticlockwise, we wandered through a patch of open, flowery woodland that would have been very lovely had the sun been shining, and here saw the first dragon of the day. It was a medium-sized something, colourless in the grey light, that skimmed unenthusiastically along through the drizzle and settled high in an elder tree. We all shuffled around the base of the tree, peering upwards, until we had a clear view of most of the underside of it. This is a view that you tend not to find in even the best field guides, so there was some debate over the beast's identity. 'I think it's a hawker,' said Graham. Shane, trying to line up a photo, sounded doubtful. 'Isn't it a bit too small?' 'Oh, then maybe it's a darter,' said Graham. Both men turned to me expectantly, two smiling, trusting, bespectacled faces. They knew — all my friends knew — that I'd gone Odonata-mad these last two years, so it was not unreasonable that they expected me to be able to put a name to this dragonfly. But I couldn't. It looked small like a darter, but showed long, delicate, leaf-shaped claspers at the tip of its rear end, like a hawker. Rather shamefacedly I told the guys I would study my photos later.

The drizzle soon stopped, and the sky began to brighten. We rejoined the main trail, heading west alongside a line of huge pylons. Lovely though Rainham is, there's no kidding yourself that you're in any kind of sweeping untouched wilderness. Luckily the wildlife doesn't mind. In fact those very pylons provide lofty perches for what is becoming a really classic cityscape bird, the Peregrine Falcon. Many cities across the UK can now boast their own breeding pair of these stunning birds of prey, the birds often improving their own PR by selecting the tallest and most spectacular landmark building as a nest site. A couple of months ago, on our Norfolk Hawker trip, Rob and I had enjoyed watching the Norwich Cathedral Peregrines wheeling around the gorgeous spire of the cathedral, over the heads of scope- and camera-toting admirers. In London itself, the pair that set up home on the Tate Modern continues to pull in crowds of tourists.

Often, a scan of the Rainham pylons will eventually reveal the stocky outline of a resting Peregrine, tiny against the massive structure. It might be digesting a recent meal or perhaps beginning to think about setting off in search of the next. Not today though. Instead, another bird of prey appeared as we were pylon searching — a Sparrowhawk that I spotted flying fast and low over the marshland. As is my habit, I talked to the hawk as I took photos of it, saying 'Come on Sparrowhawk, come back this way, please,' and amazingly enough the hawk actually did just that, banking around and coming back to soar directly over our heads. We enjoyed lovely views of it crossing a patch of blue sky, light shining through the spread of its banded flight feathers.

Sunshine brought out the dragonflies. Both of the common red darters were here, the Common and Ruddy Darters. None of us would have trusted ourselves to identify them in flight, but darters settle frequently and when perched will often allow the curious Odonata-phile to get very close to them, certainly close enough to detect the presence or absence of yellowish stripes on the dark legs. Stripe present = Common Darter, no stripe = Ruddy Darter. The little dragons liked to settle on the bare wood of the boardwalks and handrails, and both Graham and I knelt down and then lay full-length on the ground to photograph one particularly smart-looking individual, while Shane stood back, ready to warn any approaching walkers of the trip hazard we presented.

The leg colour point is the most reliable way to tell these two darters apart, something the Odonata enthusiast will need to do often as both species are quite common, are on the wing at the same time of year and live in similar habitats. However, nature does like to throw a bit of a spanner in the works and Common Darters become darker with age, including on their legs, meaning that those neat yellow identifying stripes become less and less obvious. Luckily there are a few other points to look for. The male Ruddy, as well as being a deeper, more saturated shade of crimson, most notably on his face, has a more hourglass-shaped figure than the male Common. Both sexes of Ruddy Darter have more black on their faces than do Commons.

It was encouraging seeing all the darters around, but there was, so far, a disappointing shortage of damselflies. Although the sun was now fully out, there was still a brisk breeze, enough, I thought, to discourage delicate damsels from taking to the air. I peered into every sheltered ditch and pool that we passed, scouring all floating and emergent vegetation for signs of filmy-winged life. These efforts were eventually rewarded when I spotted a very lovely blue-morph female Common Blue Damselfly crouched along a fat violet plant stem that set off her own delicate blue tones quite beautifully. Not the species I was hoping to see, but a very encouraging sign.

By the second half of August, the Azure Damselfly and the other, more uncommon *Coenagrion* species have more or less finished their flight seasons, so any mostly blue damsels you see in late summer are likely to be Common Blues. Damselfly numbers start to fall off at the same time that the commoner dragonflies are coming out in force, and I wondered if there might be a connection here. Dragonflies show no mercy towards their smaller damsel cousins, after all. The swarms of mating Azure Damselflies that throng lake shores in June would make easy pickings for the common large hawker species, the Brown, Southern and Migrant Hawkers, but these dragons aren't really flying in numbers until August. Common Blue Damselflies do share airspace with the big hawkers and regularly get eaten by them. However, Common Blues don't seem to go in for the mass-participation romps favoured by the Azures but conduct their personal lives a little more discreetly, perhaps giving them some protection against the hawkers.

At the far end of the reserve stands an impressive new two-floor hide, big enough for dozens of birdwatchers, overlooking a series of small floods and patches of rough grassland. Here, wildfowl congregate in winter, but the scene today was pretty quiet, and our visit was brief. We spent much longer by a clump of buddleia bushes behind the hide, where numerous butterflies were gathering to feed on the purple blooms. The breeze was a problem, photography-wise, swishing the flowers about and making it difficult to line up any shots. The butterflies coped with it better than we did, hanging on and feeding away as their perches were tipped, twisted and turned about.

Three butterfly species were present, all of them members of the group known as 'aristocrats', so called for their size and opulent colours and patterns. There were Red Admirals, Peacocks and Small Tortoiseshells, all strutting about on their four spiky legs … wait, four legs? Well, everyone knows that insects have six legs. But in the case of the aristocrats, only four legs are evident, the front pair being just tiny, bristly stubs, tucked up under the butterfly equivalent of a chin. They kept their balance with much flickering of their resplendent wings, in each case gloriously coloured on the upperside and plain blackish-brown on the underside. All three of these species spend the winter months hibernating in their adult form, where the closed wings afford camouflage for four months or more. Before that, though, the butterflies have to store some fat for the long fast, so they feed constantly, jostling each other for prime position on the best flowers. Come spring and their emergence from hibernation, these same butterflies that competed for resources will be more or less uninterested in feeding but instead will be hell-bent on mating.

The trail continued in an arc around the far end of the reserve, past one of the site's most striking features – the shooting range. A stepped brick bank, topped with a row huge metal numbers from one to eight, this landmark stands against the sky, elevated against the flat fields and marshland all around, and walking past it inspires a mild frisson of anxiety if you think too much about what it would have been like here a few decades ago. Looking the other way, the visitor centre is the dominant feature, a square-cut lump with its two tall roof funnels and orange-toned panelling. And close at hand is the 'dragonfly pool' – a sheltered patch of reedy, sedgey water that was, today, more or less devoid of dragonflies.

We were on the home stretch now, and took a look inside the so-called Marshland Discovery Zone, a large hide with floor-to-ceiling windows. Outside, on the boarded area right of the building, a large glass tank full of vegetation and, no doubt, small water beasties, sat centre stage. A pond-dipping session was underway here, a few schoolchildren busy at the tank with their miniature fishing nets, and I went over for a look at the contents of one of the shallow dishes. The dish contained two delicate little creatures that I recognised as damselfly nymphs.

There's no point denying it – my exploration of the world of dragonflies and damselflies has told only half the story – maybe less than half. It's understandable enough, I think, that I have concentrated on the winged, aerial, colourful life stage of the insects, because of the practical difficulties in observing them in their aquatic infancy. But the adult stage of life is a blink of the eye compared to the months or years a dragon or damsel spends as a nymph. For example, the Azure Hawker lives through three winters before it reaches adulthood, dwelling in some dark and chilly peat-bog pool. It passes the warmer months rushing and swimming after other little water creatures, eating lots of them and slowly growing, bursting out of its skin periodically when things get too snug. The final skin-ripping occurs when the nymph is full-grown, and has climbed clear of the water. What emerges this time will not be a slightly larger version of the nymph, but a full-formed, air-breathing, winged dragonfly, with no more than a few weeks of hectic adult life ahead of it.

The two damselfly nymphs swimming about in the dish were very small, probably hatched that year and with one winter to live through before they reached maturity. Their front ends were unmistakably damselfly-esque, with wide dumbbell heads and six long legs. At the back end, things were a little different. The abdomens were shorter and more tapered than those of adult damselflies, and at their tips bore three leaf-shaped fins. Both fins and abdomen moved side-to-side in a sinuous, snake-like motion. When I peered very closely at the nymphs I could make out tiny wing sheaths, held flat along the top of the back. I was considering taking out the macro lens for a photo when one of the children whipped the dish away, sending me an unimpressed glance as he did so, and returned the nymphs to their tank.

Shane and Graham, meanwhile, had wandered off to the ditch that runs alongside the trail and were busy admiring a family of Little Grebes there. I went to join them, and the three of us were soon cooing like doting grandparents over the fluffy chicks. These babies were charm personified. Dark and stripy dollops of fuzz, they regarded us more or less fearlessly with melting dark eyes, above which were fine pale 'eyebrow' markings, set at such an angle as to give their faces a permanently quizzical expression. They were neither the first nor the last Little Grebe family we saw but were by far the most confiding. Mum and dad grebe, though, were more cautious and kept their distance.

The end of the walk was nigh and we were still yet to see any red-eyed damselflies, small or otherwise. There was just one hide to go — another lavish work of art with a raised decking at the back serving as an exhibition area for various wildlife-themed artworks. The large windows looked out onto a shallow watery scrape and across the grasslands beyond. Graham detected a solitary wading bird creeping along by the scrape, which proved to be a Whimbrel — easily the most exciting bird we'd seen all day. It settled down for a contemplative stare into the water. A compact, mid-sized wader, it had the general look of a Curlew with brown plumage and long, down-curved bill, but the smart dark and light stripes on its crown gave it away. There were big smiles all round, and I started to feel a little less fed up about my failure to see the damselflies.

Just outside this hide, a wide wooden bridge crosses the same channel where we had watched the grebes a little further up. I leaned over the bridge, peered down into the water, saw nothing of interest. The guys did the same, one either side of me, and Graham said, 'I've just seen a damselfly. It's a red-eye'.

'Where? Where?' I squawked. Graham patiently pointed out the insect, balanced along a floating stem. It was too far down for me to determine whether it was a Small — the difference is a slight variation in the tail-tip pattern — but it was certainly a male of one of the red-eyed damselflies, with bulging dark-red eyes and a slim black body with a blue blob at either end. We all started snapping away, and soon I noticed two more, these joined in tandem. In all, there were half a dozen red-eyes, including a couple of drabber, green-eyed females. Their behaviour — resting in ones and twos on anything that was floating on the surface — was the same as that of the Red-eyed Damselflies on my local patch, but I knew that Small Red-eyed Damselflies are very similar in all respects to their big brothers. I took photos of all of them — a look at the zoomed-in images would tell me which species they were. From what I could see of the images on my camera screen, I was cautiously optimistic.

For the final part of our day none of us felt quite energetic enough to repeat the whole loop round, so we opted to walk part of the way along clockwise, then cut through the one-way gate and return along the Thames path, which falls outside the confines of the reserve. This meant re-crossing the same bridge from where we had seen the red-eyes, but now, just 30 minutes later, they had all disappeared.

The walk along the riverside was uneventful at first. The Thames here is a dauntingly huge entity, almost more sea than river — this is of course good news for the wildlife-watchers who have often enjoyed the spectacle of seabirds such as skuas and auks, and even the occasional seal and porpoise, in the river here. The view across the marshes from the raised riverbank is wonderful, and inspires wistful thoughts about what it must have been like here before the city sprawl ate up the original marshlands. Rainham Marshes is just a fragment of this habitat, and moreover one that has been more or less landscaped from scratch.

I felt torn between gratitude that this wonderful reserve exists, and sorrow that there is so little else like it lining the Thames' path out of London. I thought back to May and the successful quest for Common Clubtail along the lush banks of a quiet, pretty, countryside river. That was the Thames too, but some 112 miles inland as the water flows. Two snapshots of one river, and they could scarcely be more different in character.

My musings were interrupted when Shane and Graham called my attention to a female Kestrel, hovering high over the sloping bank on the inland side. We stopped to watch her, balanced expertly on a headwind, her wings beating fast, her fanned tail twisting about, and her neat little head tilted down and held perfectly steady, looking for signs of small, furry life on the ground below. Then she dropped down, but checked herself about three metres above ground, and flew at this height towards and then right over us, almost close enough to touch and near enough to for us to feel a punch of air from the downstroke of her wings. I learned later that this individual Kestrel, who hunts the bank every day and is completely un-fazed by humans, has become a bit of a local celebrity.

On the drive home, I studied the photos I'd taken of the red-eyes again and again. With the pairs in tandem, I couldn't get very far with identification, as the crucial tail-tip of the males was not visible, being curled under and fixed to the back of the females' heads. However, I felt almost sure that I could make out the little black cross at the abdomen tip of one male that I'd photographed while he was on his own – the diagnostic feature that meant he was a Small Red-eyed. A look on the big screen at home confirmed it. They also showed that the insect's black thorax wasn't black at all but reflected a rather amazing dark reddish gloss. I was still at a loss, though, to explain why this species has the 'second' name *viridulum*, Latin for 'green', until it occurred to me that maybe the species had been named for the female's appearance rather than the male's. Unusual, but it does occasionally happen.

As for the mystery dragonfly from the very start of the day, I made a tentative identification of Migrant Hawker, a species only just beginning to fly, but one that would soon be everywhere, and which would dominate my September dragonfly-watching.

CHAPTER 21

Aeshna grandis The Brown Hawker

ᘏᘓ ᘒᘏ

The split had spelled an end to serious dragon hunting for a while. I had no car so couldn't get around that easily, and in any case a lot of tedious real-life stuff to sort out. Rob and I met once more, for a trip to Wat Tyler Country Park in Essex where we looked for Small Red-eyed Damselfly and Scarce Emerald Damselfly. The day began cold and wet, but cleared up by late afternoon and we found a few individuals of the latter species, gingerly warming up on wet sedge stems around a narrow but very full pond.

Wat Tyler Country Park lies on marshy ground close to the town of Benfleet, which I knew only from a winter trip to see a big flock of Waxwings in a supermarket car park a few years ago — these lovely and glamorous birds have a penchant for the most unlovely, unglamorous locations. The map in the country park's visitor centre showed long trails leading through meadows and along riversides, and a few birdwatching hides overlooking shallow, reedy scrapes. But we were advised to look for Odonata around the small pools near the various visitor facilities.

Very similar to their commoner cousin, the regular Emerald Damselfly, the Scarce Emeralds were spectacular little insects. They seemed a little more sturdily built compared to how I remembered Emeralds to be, but had the same beautiful gleaming bodies, a metallic paint-job in British Racing Green, while the males also sported powder-blue pruinescence at the abdomen base and tip. The extent of this pruinescence is a key feature for identification — the Scarce Emerald has less of it than the Emerald. Like other *Lestes* species the Scarce Emerald tends to rest with its wings held out at about 45 degrees to the body, which creates a pleasing outline and also gives an unimpeded view of the lovely colours. It seems likely that the damselflies do this to impress each other rather than passing photographers but I wasn't complaining.

The Scarce Emerald's species name, *dryas*, comes from the Greek word 'dryad', or 'wood nymph'. This is a nice counterpart to the Red-eyed Damselfly whose species name *najas* comes from 'naiad' — a water nymph. In Greek mythology, both varieties of nymph were (very) minor deities, represented by lovely young women, and it's easy to see how a dainty, colourful damselfly might put a romantically inclined naturalist in mind of a pretty girl with a bit of magic about her. However, unlike its mythological namesake, the Scarce Emerald has no particular affinity with trees, but actually prefers more open and marshy places.

The search for Small Red-eyed Damselfly that day was unsuccessful. We found a promising-looking pond adorned with lily pads — like Red-eyed Damselflies, the Small versions enjoy sitting on lily pads — but the only damsel we found there was a very fresh Blue-tailed, boggle-eyed, frail and drab brown, its tail not yet blue but a sort of dingy lilac. Tiny midges swarmed over the surface and occasionally fell in. In the water, half-grown immature newts (newtpoles), long on tail and short on legs, rose to the surface for a mouthful of air and, sometimes, a mouthful of drowning midge. A fat male Broad-bodied Chaser came rattling by, magnificent in his fully mature, powder blue colouring. He settled on a high branch, too high for a good photo, but the perfect spot to look out for feeding and romancing opportunities.

I wasn't too disappointed to miss the Small Red-eyed. It has a long flight season, occurs at other sites nearby, and I thought there was still a fair chance of finding it at another time. As chapter 20 relates, I did find it in the end, on a trip to RSPB Rainham Marshes. But other dragon-watching opportunities were thin on the ground.

My temporary new home in East Sutton, near Maidstone, was far from prime dragonfly habitat, being set among rolling Wealdon hills with little in the way of fresh water. I had some respite in the form of a wonderful day at Dungeness (see Chapter 19), but apart from that my days were almost free of dragons and damsels. I walked the fields anyway, and delighted in sightings of Badger, Little Owl and Weasel, but my local wildlife encounters included only one memorable meeting with a dragonfly.

The dragon in question was an immature Southern Hawker, more or less fresh from the water and full of the joys of early summer. Like many dragonflies, in its early life the Southern Hawker seems to steer well away from water and finds more open and drier areas to hunt. This one was patrolling the tiniest country lane, a strip of cracked and grassy tarmac crowded in with tall hawthorn hedges. This lane was barely the width of a Mini, and few drivers dared to take it on, meaning it was always beautifully quiet. I knew the lane well, it was my favourite part of the long walk from East Sutton to the nearest train station, and I always dawdled along this part of the walk, but the hawker stopped me in my tracks.

Some dragonflies do all they can to avoid an approaching human, and you need real stealth to get near them. Others seem indifferent, and others still are curious but wary. The Southern Hawker is in another category again — very curious and completely fearless. With its breathtaking aerial ability there's little for it to fear — it could fly rings around any human, and that is what this one did to me. It 'buzzed' me at head-height, wings a blur and the graceful length of black, green-spotted body perfectly balanced between them, giving me a perfect snapshot view as it passed my eyes just a few centimetres away. It went by, then executed a full turn on the spot with balletic grace and did it again from the other direction. I registered the gleam of teardrop-shaped eyes in transition from green to blue, and as it passed the short, fat antehumeral stripes, green on a black thorax, that help identify the species. It buzzed me twice more, and then lost interest and headed away down the lane in pursuit of some unlucky fly.

That turned out to be my last walk down the tiny lane, as I got word that the tenant in the little flat I own in Sevenoaks was able to move out sooner than expected, meaning I could move back. A couple of weeks of chaos ensued, as I moved myself and my stuff back to the flat where I'd last lived back in 2008. Compared to Rob's flat, mine is tiny, but I felt a proud glow of ownership as shelves went up, belongings found their home, and things started to take shape. After an exhausting Friday working on the flat, I decided to take an evening walk down the road to my old local patch, Sevenoaks Wildlife Reserve, and see what Odonata I could find.

After such a cold, wet and uninviting July, early August was producing some true summer weather at last, and it was a warm and sunny evening as I walked down the track into the reserve. As the track opened out I saw my first dragonfly of the day and it was a beauty — a Brown Hawker chasing prey a few metres up, scything to and fro between the lines of trees with imperious, careless grace.

For many people in Britain, especially in the south, August is *the* dragonfly month. While dragons and damsels of various kinds can be found on the wing in most areas from early April until late October at least, the ones that are most likely to draw attention to themselves tend to reach their peak numbers in August. They are the common species of large hawkers — the Brown Hawker, the Southern Hawker and the Migrant Hawker. All are big, extremely active and quite happy to skim fast and loud past people's noses in a way that either provokes wonder, alarm, or both.

As dragonflies go, the Brown Hawker is not a colourful creature. Both sexes are mostly dark chocolate brown. Males have a row of small blue spots down the abdomen sides, and also a splash of blue on otherwise amber eyes, but the stockier females lack even this. However, the wings have a strong golden tint, which is noticeable at long range and makes for easy identification when the insect is cruising high against a bright sky and no colour detail on the body can be seen.

This hawker's species name, *grandis*, reflects the fact that it is one of the biggest of all our dragonflies. Only the Emperor and Golden-ringed Dragonflies have a couple of millimetres more abdomen length. Watching it in hunting mode put me in mind of something larger still — a Sparrow-hawk, another elegant hunter with a habit of skimming along lines of trees on fast-flickering wings.

I watched the Brown Hawker for a while, and then noticed there was a second one, patrolling the same area. The first went up high in pursuit of some small flying thing, only to be diverted from its target by the other hawker, which flew at its rival and steered it away from its prey. The little fly escaped while the two great dragons settled their differences with a fast and furious chase 10 metres above ground level, then they separated and resumed their up-and-down hunting flight.

I walked on, taking the trail towards Long Lake where I hoped to find damselflies, and on the way saw another four or five Brown Hawkers. Heading towards me against a leafy backdrop, one of these dragons looked at first glance like a small woodland bird with its size, brownness and fast-flitting flight. But no little bird would fly with such command-ing directness, and such predatory presence. Much as I would have liked a photo, I knew from my experiences of this species last year that it is excep-tionally restless and, if it does settle, is hyper-alert to movement nearby. So I didn't even bother to take the camera out of the bag, but just enjoyed the dragons as they, in turn, seemed to be enjoying themselves. Good hunting on a sunny evening — this disappointing year seemed finally to be working out well for the dragonflies of late summer.

All was quiet at Long Lake. Although I felt sure that there were still damselflies about, it was close to 6pm, getting towards their bedtime. They would have gone to roost, tucking themselves up among clumps of sedge and reed, positioning themselves along their slender perches so from most angles they were invisible. I did see a single Blue-tailed, freshly emerged by the look of it, flitting slowly just above the deep, black water. It moved la-zily, in its way as impressive a flyer as its big hawker cousins, its thread of a body bobbing forwards, backwards and sideways as it explored the new world on the other side of the divide between air and underwater. It hung just out of reach of the many pond skaters that walked and drifted on the water, their feet dimpling the still surface as if it were a film of rubber.

I began to make my way back, marvelling at how the vegetation had shot up since I was last here. Then I pulled up short at the intersection of three paths, where a female Southern Hawker was rushing about, presumably on a hunting mission. Curiously, she was doing this at a very low level, no more than 20cm above the ground, and was manoeuvring with terrific precision and agility in the small triangle of space where the paths met.

Back in the introduction to this book, I described a memorable encounter from years ago, with a dragonfly that was brought down by our pet cat but made a miraculous recovery. The experience did much to kick off my fascination with Odonata, and although I can never be sure I think that the dragonfly in question was probably a Southern Hawker. This dragon is a big, colourful species, with a big, colourful personality to match, and with its habit of 'buzzing' passers-by, I would guess has caused more alarm than any other British species among people who are not so keen on close dragon encounters. While out walking with a friend recently, we paused by a small pond, and a male Southern Hawker which had been patrolling the pond stopped in mid-air and hovered at eye-height for several seconds, right in front of my startled friend's face.

Male Southern Hawkers are handsome beasts, huge with vivid blue eyes, their long and slightly but clearly curved and wasp-waisted abdomens black with big beads of apple green all the way down, changing to bright blue on the last few segments. It is the males who are honoured in the species' scientific name, *Aeshna cyanea* — *cyanea* means 'deep blue'. The female of the species is stockier and has no blue on her body, just green. Both are easily distinguished from other hawkers by their antehumeral stripes, the two green markings on the top of the black thorax, which in this species are not so much stripes as near circular blobs.

Southern Hawkers aren't common on my local patch, so I made the most of watching this female as she hawked about. Both sexes of the species tend to head away from water as their first act as flying creatures, and may end up a long away from potential breeding grounds, visiting woodland glades, open heath and gardens where they hunt and mature for a few days. When they are ready to breed, the males set up territories by lakes and ponds, which the females visit first to hook up with a male, and then later to lay their eggs. I was intrigued to read of females sometimes egg-laying on dry(ish) land rather than into water, and even of one who thought the best place to deposit her precious eggs was in someone's shoe.

It was, by now, getting close to dusk, but the hawker hunted on. I had to leave her to it, but as I walked on she continued to loop around at low level, and even when I turned back and couldn't see her any more I could still hear her restless wings, a soft whirr carrying far into the still evening.

CHAPTER 22

Porzana pusilla Baillon's Crake!

❧ ☙

It was early September, and I was at home one evening, talking with Rob online about nothing in particular and simultaneously browsing the RSPB community pages. I noticed a blog post from Howard Vaughan of Rainham Marshes entitled 'Blue-eyed Beauty', and opened it up for a look. The following snippet of conversation ensued.

Marianne: OMG. Look.
http://www.rspb.org.uk/community/placestovisit/
rainhammarshes/b/rainhammarshes-blog/archive/2012/09/05/
special-hawker.aspx
I have to go and have a look for that.
Robert: Dare I suggest, weekend?
Marianne: Not sure I can wait that long!
Although it's only two days away and I'm waiting in for my
sofa tomorrow.
Robert: Well I'm up for it if you are — it's gorgeous.
Marianne: It's a stunner. Would be great to round the book
off with a rarity. As there's only been one seen, and dragonflies aren't noted for staying alive/in the same place
for ages I will try on Thursday and/or Friday and I'll
keep you posted

The blue-eyed beauty was a male Southern Migrant Hawker, a very rare vagrant dragonfly from the Mediterranean area. It is a close relative of the common Migrant Hawker, and is considerably bluer, most notably in the eye department. This dragon is bluer of eye than even the Azure Hawker. There was a photo of it on the blog post — it was certainly a beauty, and I desperately wanted to see it. So did Rob, of course, but I couldn't wait for the weekend to travel there with him, time was of the essence. My long-awaited sofa was to arrive the next day (Thursday) and I decided that if it showed up early enough I would go to Rainham that day; if not, the day after.

The sofa delivery didn't happen until mid-afternoon, so Friday it was. I tried to round up some friends but no-one was free at such short notice, so I set off to the station alone, to make my way to Rainham by public transport for the first time. With the Paralympics still on and many people taking the train to Stratford, it was a busy trip, but once I disembarked at Purfleet I was on my own. I followed the river path to the reserve, seeing a couple of Migrant Hawkers on the way. For once the weather was perfect for dragonflies.

The Migrant Hawker used to live up to its name and occur only as a wanderer to the UK, from its breeding grounds further south. From the 1950s it began to establish a breeding population in the south-east, and it has gradually spread north and west, to the point where now it breeds across nearly all of England and south Wales, and is marching determinedly towards the Scottish border. Although now a well-established resident here, it does still occur as a migrant as well, meaning that in early autumn it can be amazingly abundant in southern England. It seems possible that the Southern Migrant Hawker is set to follow in its footsteps.

Going into the visitor centre, I was greeted at the desk by Howard Vaughan. A genial man, he was a dedicated birdwatcher at the marsh well before the RSPB took over, and they had the good sense to take him on as reserve manager. He is a tireless ambassador for wildlife, and a real star of this flagship reserve. I asked him for advice concerning the Southern Migrant Hawker, and he told me where it had been seen (by the dragonfly pool, strangely enough), and showed me a print-out he had made up with side-by-side photos of Migrant and Southern Migrant Hawkers, highlighting the main differences between the two species. As I thanked him and turned to go, he called after me, 'Check every hawker!'

It became apparent very quickly that checking every hawker, while certainly desirable, was going to be incredibly difficult. Since my last visit just a couple of weeks ago, the place had turned into Migrant Hawker City. Everywhere I looked, there were several of the blue dragons gliding elegantly past. They loitered low over water in the ditches and sped across the rustling tops of the reedbeds. If I looked up I'd see a few high-flying after some prey or other, silhouettes against the blue sky. I'd barely made a start on the trail and I had already seen a couple of dozen of them, and only positively identified about half of those.

I carried on, making my way clockwise around the path, as this would get me to the dragonfly pool a bit more quickly. All around me, the Migrant Hawkers hovered and dashed. When I paused on a bridge over a creek where I've sometimes seen Water Voles, there were half a dozen Migrants playing around on the water's edge, and a couple of them settled and allowed me to take some photos.

All of the Migrants I had seen so far were males. They are showy dragons, on the small side for hawkers, black-bodied with quite prominent blue blobs down their bodies' length. They are sometimes confused with Common Hawkers, but lack that species' super-slim wasp-waist, and have much reduced antehumeral stripes — just two little yellow spots on the thorax, while the Common sports full-length yellow stripes. The Migrant Hawker also has a distinctive 'golf-tee' marking on its second abdominal segment. Its eyes are bright blue at the back edge, changing to blackish at the front. This means that a Migrant flying away from you looks as though it has eyes as blue as the blue-eyed beauty itself, but when it swings about the two-tone pattern is obvious. *If* it swings around.

Maybe the potential for confusion with other species is why the Migrant Hawker's species name is *mixta*, which means (as you'd think) 'mixed-up' or 'confused' in Latin. Or maybe that refers to something different. The kaleidoscope of colours in its eyes? Its behavioural quirk of hovering on the spot before quickly changing direction, as if it kept forgetting where it was going? Its tendency to migrate long distances away from where it was 'supposed' to be? I can't find a definitive answer, so I guess it can mean whatever you want it to mean.

The whereabouts of the female Migrants was a mystery. I saw a few, maybe one for every 30 males, and each female I saw was firmly attached to a male in a mating wheel. I managed a few photos of one such pair on a sedge leaf before someone coming the other way scared them off. The lady half of this pair was a much duller specimen than her partner, with a brownish black body marked with small yellow spots, and no hint of the vivid blue that adorned the male. Her lack of showy colour didn't make her any less attractive to him, and no doubt he was the envy of all the unpaired males zooming around.

Most dragonflies and damselflies have obvious sexual dimorphism — in other words, the males and females look distinctly different — and it is the male of the species that has the showy colours. Females are drab and dull for good reason. They have the risky job of egg-laying, which usually requires them to be still for relatively long periods on or even in the water, a vulnerable position for any insect. Camouflage helps reduce their chances of being snatched by a predator at this crucial time.

The females don't just dress modestly; their behaviour is more discreet and furtive too. Often females will spend most of their lives away from water, only going there when they want to mate. There's no need for them to draw attention to themselves when they approach — each patrolling male is watching out for a passing female. He will quickly waylay any female that comes along and try to grab her body with his legs and lock onto her head with his claspers as quickly as possible, before another male can get in on the act. It is all rather high-speed and brutal-looking, but to complete copulation the female does need to co-operate and curl her abdomen forwards to meet his. If she is not interested she flies fast to try to dislodge him and curves her abdomen away rather than forwards.

The dozens of male Migrant Hawkers were all on the hunt for female company, but they were also looking out for each other. Each patrols a waterside territory — a spot that will attract females — and so needs to be highly visible to keep other males from encroaching on that territory. But really these males are lovers, not fighters. Not for them the violent scrapping of species such as the emeralds and chasers. Perhaps they refrain from combat because there is too great a risk of a female slipping past them while they are otherwise engaged.

I reached the dragonfly pool, under the looming wall and mounted metal numbers of the firing range. A boardwalk traverses the well-vegetated pond, and here I loitered, watching lots of Migrant Hawkers going about their business. Among all these big blue dragons flew a few small red ones — Common Darters. Maybe there were a few Ruddy Darters too, but today wasn't a day for examining darters' legs. I was on high alert for Old Blue Eyes. But after half an hour loitering by the dragonfly pool I had nothing to show for it.

I decided to move on. The Southern Migrant Hawker would, no doubt, have flown around pretty freely so there was every chance it had moved away from the pool. Or gone completely — but I wasn't ready to give in to such negative thoughts, not just yet anyway. I walked on to a spot where the boardwalk crosses some open water and sat down here to see if any hawkers came along. At the very least I hoped to get some nice flight photos of the Migrant Hawkers, which should have been doable as they are quite prone to hovering on the spot.

I'd been sitting for just a couple of minutes when I heard a scrunching noise among the sedges that lined the pool. I lifted my binoculars and pointed them in the direction of the sound, and after a moment of squinting I discerned a curious thing – a little bright orange object that seemed to be moving up and down a stem. I stared in confusion at the object, and eventually figured out what it was – the incisor teeth of a Water Vole, which had climbed up into the vegetation to access the bits it particularly wanted to munch. The animal's little face was hard to make out in the shadows but as its head moved up and down I saw the occasional sparkle of a little polished-button eye. I waited, hopeful that the little mammal would come out into full view, but it remained in hiding, and when some birdwatchers came along it quietly disappeared.

The Water Voles at Rainham are thriving, in contrast to the general picture nationally for these charming little creatures. This is thanks to the very extensive waterway creation and reshaping work that's gone into the reserve over the years, to create ideal habitat for Ratty. The quiet, lush ditches in particular appeal to the voles, and if you wait patiently somewhere with a good view across a ditch there's a good chance that you'll see one paddling steadily across from shore to shore, little blunt muzzle pointed skywards. If you walk about quickly and impatiently, you're more likely to hear the distinctive neat 'plop' as an alarmed vole leaps into the water at the sound of your stomping feet.

Volunteers from the London Water Vole project have been monitoring the voles since 2001, and have recorded a steady increase at Rainham. There is hope for Water Voles nationally as well, curiously because of the resurgence of another charismatic water mammal. Otters are doing well almost everywhere, and as they advance, the numbers of the non-native American Mink – arch-enemy of Water Voles – are falling. Although Otters probably don't directly kill minks (or at least don't do this habitually) their presence is enough to force the minks to move on, and as Otters take over the best riverside habitats the minks run out of room and cannot breed successfully. More Otters, fewer minks and more Water Voles is the right way to go for better and more balanced wetland ecosystems, as far as conservationists are concerned.

The other animal in evidence at this spot was the Marsh Frog. From where I was sitting I could see three, the bulgy tops of their heads just showing through a film of waterweed close to the pool's margin. The Marsh Frog is a large and loud amphibian, handsomely marked in shades of vivid green, and unlike the Water Vole it is not native to Britain. According to the Surrey Amphibian and Reptile Group, its presence here dates back to 1935, when a Mrs E. P. Smith of Ashford, Kent, decided to treat her husband by stocking the garden pond with Edible Frogs from France. As genuine Edible Frogs weren't available, she obtained a stock of Marsh Frogs from Hungary instead. Not wanting to hang around and possibly be eaten, the frogs left the Smiths' pond and indeed garden, and set up home in nearby ditches. From here they have spread along the network of waterways across south-east England and become firmly established. Their impact on native ecology is still not clear, though they have certainly provided a welcome addition to the menu for some of our predators, like Grey Herons that are only too happy to gulp down even the most monstrous Marsh Frog.

The species' scientific name is *Pelophylax ridibundus*. I had to look up *Pelophylax* (it's a combination of the Greek words 'mud' and 'guardian', an apt enough description of this frog's fondness for very wet places). But *ridibundus* I knew already — it means 'laughing' and describes the frog's crazy call. People walking in areas full of Marsh Frogs are often taken aback by a sudden chorus of noisy, croaky laughter from the ponds or creeks as they pass. One frog beginning to call sets off all the others, and they can keep going for many minutes, before all suddenly go quiet again. If you get a good look at a Marsh Frog in mid-guffaw, you'll see the translucent grey bubbles of its vocal sacs inflating and deflating on either side of its cheerful face.

I watched one of the Marsh Frogs swim lazily along the edge of the pool, to join another. A blue dragonfly appeared and hovered briefly (too briefly for a photo) over the corner of the pool — it was, of course, yet another male Migrant Hawker. I stood up and walked slowly onwards towards the two-storey Tower Butts hide, wondering if I could be bothered to complete the whole loop, or whether it might be better to turn around and head back for the dragonfly pool.

At this point, I met a birdwatcher heading the other way, who told me that there were several Hobbies showing well from the Tower Butts hide. This news convinced me that I should at least spend some time in this hide, for I enjoy watching these lovely little birds of prey as much as the next person. So I went into the big, empty hide and sat down in a corner seat to scan the pools and grassland ahead.

I spotted a Hobby straight away. There was no mistaking that slim, sickle-winged shape and effortless wheeling flight. It was having a field day chasing dragonflies – as I watched, it snagged a dragon in one foot while already clutching one in the other. Could I be looking at the reason why I hadn't managed to find the Southern Migrant Hawker? Looking further I found two more Hobbies, also engaged in dragon-slaying. Lovely though they were, I couldn't help feeling a touch of bitterness towards them – it seemed most unlikely that they would have the sensitivity to pass over the rare dragon and stick to only eating the common ones.

There wasn't too much else to see from the hide. On the strip of water nearest to the hide were a Little Grebe family, the chicks half-grown balls of stripy fluff, the sleeker but harassed-looking parents diving repeatedly and surfacing with a little wriggling fish or insect larva to pass to one of their squeaking offspring. A distant Little Egret, angelically white and graceful, flapped across the fields to pitch down unseen in a flooded ditch.

I stayed in the hide longer than I should have. I knew I ought be getting back out there and putting in a last hour's search for the Southern Migrant Hawker. But there was something hypnotic about those graceful Hobbies as they deftly and elegantly munched their way through Rainham's dragonfly population. It was also a bit of a novelty to have this spacious hide all to myself. I decided I'd procrastinate just a little longer, and aimed my camera at the Little Grebe family, to take some photos that I knew wouldn't be up to much, the birds were just too far away.

It was at this point that the day took a very strange turn. Behind the grebes, in the thick vegetation lining the water's edge, I noticed a movement, and raised the lens to focus on whatever it was. I registered a smallish, squattish something, with vertical black-and-white barring on its flanks. Even though it was really tiny in the viewfinder, I took a few shots as it picked its way along the edge and then slipped deeper into the vegetation, out of sight.

I hit the preview button and zoomed in on the camera display. There on screen was a greyish, brownish bird, with a rather short bill and long legs, the aforementioned barred flanks, and white speckles across its wing feathers. I realised at once that it was a crake of some sort, and felt a frisson of excitement. I looked along the waterside to see if it had reappeared. It hadn't. I went back to the photo.

I could immediately narrow down the field to three suspects, one scarce and two very rare. The scarce one is Spotted Crake, which has a small breeding population in Britain and also visits on migration in similarly small numbers. The very rare ones are Little and Baillon's Crakes, both with fewer than 100 records in Britain. Which one of them was this? I didn't have a field guide with me so couldn't check the crucial plumage features. It was a test of memory. Baillon's Crake and Little Crake were similar, I knew, both markedly smaller than Spotted Crake and with similar plumage. I have seen Little Crake before, in 1997 at Bough Beech Reservoir in Kent. My memories of that bird were that it was tiny, but the bird I'd just photographed hadn't struck me as tiny, merely 'smallish', and it looked quite round in the photos. And it had white spots. And Spotted Crake was the likeliest option of the three, numbers-wise. I decided I had probably seen and photographed a Spotted Crake, but I knew that I needed to check my photos against a field guide, and I also knew that I needed to tell Howard or someone else at the visitor centre about my find, because this would be a bird that other people would want to see.

After a further short wait to see if my crake reappeared (it didn't), I left the hide and began to walk back to the visitor centre. I felt torn between wanting to run all the way and wanting to make the most of my remaining time on the reserve on such a lovely day. In the end I did stop for a few minutes at a waterside spot to exploit the chance to photograph an obliging Migrant Hawker, and also a Marsh Frog swimming through beautifully clear water. A little further down the track, another un-missable photo opportunity came my way – a female Kestrel sitting on a fence post, peacefully cleaning her breast feathers of the gore from her latest kill. This bird, who is often encountered on this stretch of path and who has been a frequent star of Howard's blog, seems to have no fear of people and let me get close enough for some frame-filling pictures. I wondered if it was she who had skimmed the top of my head when I'd been here with Shane and Graham a couple of weeks before.

Finally I was back at the stairs leading up to the top floor of the visitor centre, and there by the little RSPB shop was Howard, who asked how my hawker hunt had gone. I explained that I'd had no joy, and then said, 'but I did find something else that may be of interest.' His eyes lit up with curiosity as I faffed about finding a crake photo, then shoved the camera under his nose.

His reaction was a sight to behold. I am fairly sure he stopped breathing for a little while. Then he turned to me with an expression frozen somewhere between delight and amazement. 'Hang on. Don't go ANYWHERE' he said, and hurried into the RSPB shop, returning moments later with a copy of the *Collins Bird Guide*, essential tool for every British birdwatcher. I held the camera and kept refreshing the image while he flicked through the pages to Rallidae. 'How big was it?' he asked as he compared my photo to a page full of crakes. 'About Spotted Crake size,' I replied, giving away what I thought the bird was. But then Howard amazed me by saying, 'I think that's a juvenile Baillon's Crake.' Then he asked me if I knew what has been going on with Baillon's Crakes this year.

What a question. And it's hard to express just how ashamed I felt to admit that no, I didn't know what had been going on with Baillon's Crakes this year. In any other year, I'm sure I would have — I am a fairly committed birder and do normally keep my ear to the ground with regard to significant birdy news. This year I've been a bit distracted though — the dragons and damsels are to blame. How strange it was that on my last major dragonfly trip of the year I should be drawn back into the world of serious birdwatching.

Howard told me the Baillon's Crake story. There had been a national survey looking for Spotted Crakes, to try to map the breeding distribution of this rare bird. Because Spotted Crakes are amazingly shy birds, the survey involved making sound recordings well after dark, as the male Spotted Crake likes to give its territorial whip-cracking 'song' by night. At a few of the survey sites, though, surprise guest singers had joined in — male Baillon's Crakes. Prior to this, it had been thought that Baillon's Crakes, though present in parts of western Europe, did not breed in Britain, but all of a sudden there was this quite compelling evidence that, actually, they did. Rainham Marshes had not been surveyed — on the only nights when there were people available to do a survey, there had been horrible weather that would have silenced even the most amorous crake. But the habitat at Rainham was said to look very Baillon's-friendly, according to researchers who had studied the species in the Netherlands.

Howard asked to borrow my camera's memory card so he could copy the photographs. I said yes, of course, but I wasn't about to let the card out of my sight so I went upstairs with him into the buzzing hub of an office. People looked up in surprise from their workstations as I walked in, and I decided to compound my cheekiness by refilling my water bottle at their kitchenette tap while Howard loaded up the photos onto his PC.

Looking at the crake pictures on screen, with the field guide to hand, it was obvious that Howard's identification was correct. What did this mean? First of all, there was the exciting possibility that this extremely rare little bird had been born and raised on the reserve. It was a juvenile after all, although young crakes may begin to migrate (they overwinter in Africa) while still in juvenile plumage so we couldn't be sure. The idea that Baillon's Crakes had been here all spring and summer, nesting and rearing a family and going completely undetected by the many keen birdwatchers visiting Rainham every day was somewhat mind-boggling. But these birds are so shy and discreet that it's perfectly possible that this was exactly what had happened.

As Howard copied the photos, he made a remark to me that I had just got myself in the record books. This was a strange thought. A few years ago, I had worked on the magazine *Birdwatch*, and one of our regular 'things' was to publish reports from the people who'd found the rarest birds in Britain that month. I wondered if my old workmates would be approaching me to write up my sighting.

Howard was talking about 'putting the news out'. There was no doubt that birdwatchers would be eager to try to see the crake, so Howard would need to notify the various rare bird alerting services, who keep their subscribers abreast of the latest rarity news via pager messages. He spoke of needing to open the reserve at dawn and close at dusk, rather than the usual hours of 9.30am to 5pm, to accommodate the hordes of twitchers who would come to see the crake. A busy day, possibly many busy days, lay ahead — he was starting to look exhausted just talking about it. I needed to leave him to make the arrangements.

Over the next few days, it was good news for the twitchers as 'my' Baillon's Crake was seen again, several times. Some of those who came to see it had to wait several hours but nearly everyone eventually went home happy. I had my 15 minutes of fame with a short piece in *Birdwatch*, and I weighed in on a few Internet message boards to quell the rampant Chinese rumours about the circumstances of my find, in which I had somehow turned into a clueless non-birder who thought she'd photographed a Water Rail. The crake was last seen on 23 September, so it stayed for 17 days in all before moving on, and in that time several hundred twitchers made the trip to see it. Birding websites filled up with photos and video footage of the little bird going about its business, along with shots of the inside of the Tower Butts hide, crammed to capacity with people and telescopes.

But I'm getting ahead of myself. I left Howard to spread the word and prepare for an onslaught of twitchers, and walked back to Purfleet station alone, in the quiet of the late afternoon. I felt an odd mixture of elation and disappointment, and asked myself whether I would have swapped the finding of the crake for a look at that Southern Migrant Hawker. It will probably cost me what's left of my birding credentials to admit it, but yes, I think I would. After two summers chasing dragons and damsels, these beautiful and wonderfully charismatic insects have won themselves a very significant place in my heart, and I would have absolutely loved to have rounded off the summer the same way the quest began back in April 2011 — with the sighting of a spectacular rarity.

Life doesn't always deliver what you want it to, though. And as I headed home on the train, I contemplated the winter months ahead, devoid of dragons and damsels. They would still exist of course, but as nymphs, dwelling in their murky and mysterious underwater world away from curious eyes and camera lenses. Maybe my find of the crake was a little reminder that (wild)life goes on, and when the dragons and damsels are gone there will always be birds to watch. After all I was, first and foremost, a birder. But I knew that, even though the stories for this book would end here, with one that got away, my passion for the Odonata would stay with me for life.

CHAPTER 23

The Ones That Got Away

ॐ ॐ

In birding parlance, to 'dip out' or just 'dip' means to fail to see an expected or hoped-for bird. The term doesn't really adequately convey the sometimes crushing disappointment of the experience. If someone else *does* see the bird while you don't (perhaps you were looking the wrong way at the one crucial moment), this is described by a more meaty term — the other person has 'gripped you off'.

Rob and I dipped out on two of the Odonata species we'd made special trips to see in 2012 — the Southern Damselfly and the Scarce Blue-tailed Damselfly. The sad story of these two failures is recounted in Chapter 18. Later in 2012 I dipped out on the Southern Migrant Hawker, as described in Chapter 22, and it could be said that I was also gripped off because a friend of mine actually saw it at the same site on the very next day. But there were also a few species that we didn't attempt to see, either because they only occur as wandering vagrants and didn't happen to turn up anywhere that we could get to in time, because they are very rare and sporadic breeders whose breeding grounds were out of reach to us, or (in one case) because they live an unaffordably long way away. I sincerely hope that in years to come I'll get to meet all of them, but for now I can only admire them from afar.

The genus *Lestes* (a Greek word meaning 'robber' or 'bandit') comprises the emerald damselflies, or as they are known in America, the spreadwings. Both names are apt, for these damsels have metallic green bodies, and unlike most damsels rest with their wings open rather than closed. The group includes two well-established UK damselfly species —

the Emerald Damselfly and the Scarce Emerald Damselfly, and, world-wide, many more. A couple of these species are in the very early stages of colonising the UK, most notably the **Willow Emerald**.

This is a big, long, dark green damsel, which shows much less difference colour-wise between the sexes than do most damsels (including other emeralds). Males are slimmer than females, and have blue eyes rather than brown. The Willow Emerald is quite widespread on mainland Europe and, since the early noughties, has gained a foothold in southern England, with the first evidence of its presence being an exuvia found in Kent by serious Odonata experts John and Gill Brook. Now, a few individuals are seen each year and the species is slowly spreading.

I am embarrassed that I didn't manage to see any Willow Emeralds. One of my fellow Kent wildlife bloggers, the talented photographer Marc Heath, birdwatches at Reculver on the north Kent coast. He began posting enviably wonderful Odonata photographs in his blog during the summer of 2012. In September he put up some shots of Willow Emeralds. I asked him for some info on where the damsels could be found, which he provided, and I started investigating how on earth one would get to Reculver without a car. After lengthy exploration of websites and Google Maps I concluded that it wasn't going to be possible. Well. It *would* be possible if I didn't mind walking many miles with my camera gear, from the nearest station (Herne Bay). I knew I wasn't up to the walk, and the bus service was too sporadic for a practical return trip. And life by then was very busy anyway. I had to rule it out, but at least I could enjoy Marc's photos.

Looking at the photos of a female Willow Emerald, they show a graceful dark damsel perched pertly in a typical *Lestes* stance, with abdomen at 45 degrees to her vertical perch, and her wings held at about 45 degrees to the body. This pose reveals the full length of body, which you could describe as 'green' (and the species name, *viridus*, means just that) but you'd be doing the damsel a disservice if you didn't also mention the flaring tones of gold and copper that sparked off her elegant form where the sunlight fell. On the really close-up shots, the iridescence shows as a scattering of bright glitter, and the great spheres of her eyes graduated from rich chestnut at the tops to lime green at the bottoms.

Another emerald, this one a little less dazzling, also seems to be colonising Britain. The **Southern Emerald** (species name *barbarus*, meaning 'stranger' in Latin) has a wide distribution in Europe, especially the south, and since 2002 it has been present in tiny numbers (no more than 10 seen at any one time) at a site in Norfolk. This is Winterton Dunes on the east coast, an area of open heathland that grades into the steep up-and-down rolls of a massive dune system, held together with thick, tough marram grass. The pools where the Southern Emeralds live are shallow and stagnant with little vegetation, so they are not very inviting and perhaps not very stable. The good news is that Winterton Dunes is protected as a National Nature Reserve, so even if the Southern Emerald doesn't successfully spread to other sites, it should be well looked after as long as it remains here.

Southern Emeralds are a little smaller than Willow Emeralds, and are a lighter shade of green with less metallic iridescence. Their eyes are rather pale and glassy-looking. Males have a touch of powder-blue pruinescence on the base and tip of the abdomen, though nowhere near as much as male Emerald and Scarce Emerald Damselflies. Their liking for these coastal duneland sites means that they may struggle to gain more than a toehold in England, with her mostly built-up coastline, but time will tell.

I agonised long and hard over whether to attempt to see the **Irish Damselfly**, a species related to the Azure Damselfly that occurs in Northern Ireland and the Republic. For completeness, I knew that I should — but on the other hand, the idea felt faintly ridiculous. I've never been to Ireland, have always wanted to, but ... really? A trip to Ireland, a flight and at least one night's stay, with all the costs that would entail, just in the hope of seeing one small, blue damselfly which looks very much like all the other small, blue damselflies? Of course, there are other Odonata species in Ireland, but none that I couldn't easily find elsewhere.

I decided in 2011 that I wouldn't make an attempt to see the Irish Damselfly. Then when the dragon search restarted in 2012 I had the same dilemma all over again. The 'it's only another small, blue damselfly' argument had become less persuasive. The Irish Damselfly was every bit as special and interesting a creature as every other Odonata species, it just happened to live inconveniently far away. Then I discovered that there are flights to Belfast from an airport near us whose existence I had barely registered before — London Southend. I checked out the prices — they were affordable. We could, conceivably, go and bag our damselfly and return on the same day.

In the end, though, the weather put a stop to these fledgling plans. We had the worst June and July I can remember, and although there were some good days here and there — and undoubtedly on those days there would have been Irish Damsels out and about for all to see — the risk was too high. It was one thing driving for an hour to a site when there was a chance of a few sunny intervals, but another to book a day off work, a flight and a hire car on such an uncertain outcome just wasn't justifiable.

The Irish Damselfly is a member of the genus *Coenagrion*, making it a sister species to the Azure, Variable, Southern and Northern Damselflies. Why it doesn't occur in Great Britain, only Ireland, is beyond me, as it does live on mainland Europe, and the right kind of habitat for it (small, lush and sheltered lakes) is certainly present. Its specific name, *lunulatum*, means 'little moon', and describes the crescent-shaped marking on the male's second abdominal segment — more like a boomerang than a moon to my eyes. This body part is key to identifying a male blue damselfly as each species has a different-shaped black marking set on the otherwise blue segment — but in the case of the Irish Damselfly, most of the other abdomen segments are solid black too. It's the darkest *Coenagrion* and thus probably the easiest of all to identify.

There is actually another rather dark *Coeagrion* species in Britain, but this one lives, as far as we know, at just one site in north Kent, and so is unlikely to cross paths with the Irish Damselfly any time soon. This is the **Dainty Damselfly**, and its history as a British species is a strange story. It was first discovered over here in Essex in 1946, on coastal marshland. The first sightings were of just a couple, but then a thriving colony was found at a pond near the town of Hadleigh. However, the Essex idyll proved short-lived for these little damsels — in the late winter in 1953 there were cata-strophic coastal floods all along the east coast, following storms over the North Sea. Sea defences were completely overwhelmed, and more than 300 people lost their lives in the east coast counties of Essex, Suffolk, Norfolk and Lincolnshire. Every known Dainty Damselfly site was in-undated, and no more were seen in the summer of 1953. The Dainty had become extinct in Britain, less than 10 years after it was discovered here.

Fast forward to 2010. By an obscure little pool on the Isle of Sheppey, Kent, those redoubtable Odonata experts John and Gill Brook struck it lucky again with the find of a female Dainty Damselfly. In 2011, more were present, some seen mating and egg-laying. I was keen to go, the site was within easy reach of home, but it just somehow didn't happen. Not to worry, I thought, I would go in 2012 instead. I felt quite happy in this decision, in fact was very confident that the damsels would be easier to find than before. Not everyone who went to look for them in 2011 was successful, but in 2012 there would, presumably, be more of them as the 2011 individuals had been seen engaged in breeding behaviour. But to my dismay, there were no reports of Dainties at all in 2012, despite searches by the local Odonata-philes.

It was time to kick myself, eat humble pie, and generally feel regret-ful over my failure to go and look for the Dainties back in 2011. Maybe I should have headed for Sheppey and had a look anyway — perhaps the damsels were there but no one was looking on the few good-weather days we had during their flight season. I'll be interested to see what happens in years to come — will the Dainties follow in the footsteps of the likes of Small Red-eyed Damselfly (first recorded in 1999, now present across much of England) or will they always have a marginal presence in Britain at best?

The Dainty Damselfly is, according to the book, the same size as the familiar Azure Damselfly, but gives the impression of being smaller and more delicately made, a trait expressed in its scientific name as well as the English name (*Coenagrion scitulum* — the word *scitulum* means 'pretty' or 'neat'). The male has a stalked, thick-bottomed U-shaped mark on his second abdominal segment, like a combination of the corresponding markings on Azure and Variable Damselflies. To my eyes it is the female who is prettier. Her abdomen segments look solid black at any distance but with a close view you can see that they bear tall black markings shaped like cypress trees on a blue background.

So that's a total of six damselflies which were around in 2011–2012 but which I didn't manage to see. There is a seventh species on the British list that I missed out on not by a few days or a few dozen miles, but by 54 years. The **Norfolk Damselfly**, *Coenagrion armatum*, became extinct in Britain in 1957, after the few sites where it lived in the Norfolk Broads dried out. This delicate little damsel does still exist on mainland Europe so it may yet re-colonise the UK, but in the meantime it rejoices in a singularly inappropriate English name. Curiously, its species name *armatum* means 'armed'. This possibly relates to its second segment marking, which is a sort of pointed blob that could be considered spear-like if you use a lot of imagination.

What are the chances of a re-colonising species being discovered? I would say that they are very good, as more and more people become interested in dragons and damsels. Many such Odonata enthusiasts are also keen birdwatchers, and an awful lot of very keen birdwatchers live in Norfolk, generally held to be the best birding county in England. With many southerly Odonata species beginning to show up in increasing numbers over here, and with the encouraging recent return (albeit possibly short-lived) of the Dainty Damselfly, there must be a fair chance that Norfolk Damselflies will be found back where they belong, one day.

Another damselfly species which may, or may not, be set to colonise Britain is the **Winter Damselfly**, *Sympecma fusca*. This species has, so far, made it over here only once (that we know about). Although the Winter Damselfly isn't much to look at (its colour scheme is a mixture of muddy brown and straw yellow, with the only hint of brightness being the male's dark blue eyes), it is a very interesting little damsel. The only record comes from south Wales, in 2008, so it was never on our radar as a species we might see. As the damsel hails from mainland Europe, the one found in Wales must have overflown a fair chunk of England to arrive there – unless it had some kind of human assistance – suggesting some potential for migration and perhaps therefore colonisation of new grounds. It belongs to a small genus, closely related to the emerald damselflies and it, plus one of its congeners, *Sympecma paedisca*, are the only European Odonata that overwinter in their adult form.

As for the dragonflies, I am pleased to report that I saw every single regularly breeding species, plus one very rare wanderer into the bargain. However, the Vagrant Emperor – and the one that I missed, the Southern Migrant Hawker – were not the only dragon rarities that graced British shores in 2011–2012.

The **Lesser Emperor** is rather similar to the Vagrant Emperor, and prior to 2011 was certainly the more likely of the two to make an appearance over UK soil. This is a medium-sized, slim dragon, nowhere near as big or as bright as the Emperor. The male Lesser has a dark, drab body, relieved by a blue 'saddle' at the abdomen base, just before the very pinched-in waist, and compellingly bright green eyes. The female actually has more blue on her than her mate, extending almost the full length of her abdomen (which is plumper than his), but it is a little duller.

Lesser Emperors are recorded in Britain most years, and 2011 and

2012 were no exceptions. In fact, my fellow Kent wildlife blogger Phil Sharp saw a female on his local patch near West Malling in July 2011, and was more than willing to show me to the right spot, but the pesky dragon was not relocated. Another observer on the same site had already seen a pair mating, a few days before. Phil and I were hopeful that this may mean an emergence of a new generation of Lesser Emperors in 2012, but sadly this did not come to pass. There were other records, but they were either too far away or I heard the news too late to consider a trip.

In June 2012, there was a remarkable dragonfly record in Suffolk – a **Yellow-spotted Whiteface**. Not the first ever seen in Britain, it was however the first certain sighting since 1859. Though it was around for three days, the news didn't get out onto the main Odonata grapevines until after it had left, so I was spared having to decide whether to bribe someone to drive me to Dunwich Heath to look for it. This is a great-looking dragonfly, belonging to the same genus as the White-faced Darter, and in its nearest regular breeding grounds over in the Netherlands is apparently increasing at an impressive rate so there are likely to be more records.

A few lucky people at Dunwich took excellent photos of the male darter as it basked on a wooden railing. The pictures show a small but sturdy, dark red dragon, with glossy dark blue-green eyes and a neat pale yellow blob two thirds of the way down the body length (on the seventh segment). A little further searching turned up some pictures from Europe of females, which were just as attractive as the males – browner, but with a whole row of yellow blobs rather than just the one. If this dragon does become established in Britain, I predict that we will start calling the White-faced Darter by a more Euro-friendly name of Small Whiteface, to show the two species' close relationship.

Another darter species is on my 'coulda, woulda, shoulda' list. This is the **Red-veined Darter**, so called because its wing veins are red rather than black. It is otherwise pretty similar to the Common Darter, the males red and females yellow-brown, though is a little lighter and brighter in both sexes and has a yellow flush at the hindwing bases. Its scientific species name, *fonscolombii*, honours a French entomologist who rejoiced in the name Étienne Laurent Hippolyte Boyer de Fonscolombe. After a year's incarceration in the aftermath of the French Revolution in 1793, this son of an aristocrat devoted himself to the study of insects until his death in 1853.

Red-veined Darters turn up in Britain every year, most often in the south-west, and sometimes breed for successive years, but their colonies don't seem to persist for more than a few years — it could be that truly stable colonies will be established in the very near future.

The most impressive of all the migrant dragonflies to reach Britain, in terms of distance covered, has to be the **Common Green Darner**. This is a North American dragonfly, so the only way for it to get across to us is by an Atlantic crossing. Consider that for a moment: a 7cm dragonfly, with 5cm wings, flying non-stop across more than 2,000 miles of open sea! And it wasn't just one dragonfly — in September 1998, up to nine Common Green Darners turned up, some on the Isles of Scilly and some in Cornwall.

Transatlantic vagrancy is well known among birds. Every autumn, some of the keenest British twitchers make a pilgrimage to west coast islands or headlands, such as the Isles of Scilly or the Hebrides, hoping that a few lost migrating North American birds will turn up, and every year some do. The feat of the crossing is not so remarkable for the larger, stronger-flying birds, such as gulls and waders, but when it comes to the tiny wood warblers and flycatchers, it seems nothing short of miraculous. Arrivals of these 'Yank' birds, as the twitchers call them, are often accompanied by a few Monarch butterflies, the occasional North American moth, and on the aforementioned occasion in 1998, the Common Green Darner dragonflies. When insects start joining in, then the superlatives start running out.

Of course, there's a bit more to this than a bird — or dragonfly — deciding on a whim to head east straight out to sea on an otherwise ordinary day. The species involved are on the move anyway, and many of the birds are youngsters, making their very first southwards migration. Add in a bit of inclement weather, and they can get pushed out to sea, where undoubtedly most of them will perish. But if there is a strong westerly wind, that may be enough to help propel even these small and delicate creatures all the way across. Some birds even pitch down on east-bound ships and make the journey that way, with relative ease if they are lucky enough to find a food source on board. In 1998, there were major hurricanes over in North America and as well as the darners, stray American birds like Western Sandpiper, Bobolink, Common Nighthawk and Rose-breasted Grosbeak also showed up over here.

The Common Green Darner, named for its pointy-tipped body, which vaguely resembles a darning needle, is the transatlantic version of our Emperor Dragonfly. It belongs to the same genus, *Anax*, and is rather similar in appearance with a blue body and bright green thorax in males, and everything green in females, though it is a little smaller with darker eyes. Its species name *junius* may be related to the month of June, although it actually flies from April into November in some areas. In autumn, many of the darners in the north will migrate south, sometimes at a fair height, and so are vulnerable to being caught up in whatever winds happen to be blowing.

The 1998 arrival of the darners caught the attention of some national media, and inspired a magnificently hysterical headline in the *Independent* — 'Giant bird-eating dragonflies cross the Atlantic'. Anyone not too freaked out by that to read on would have found a pretty balanced account of things, though. As for 'bird-eating', there are records of darners in America having a go at some of the tiniest hummingbird species, though I can't find any accounts of one actually consuming this unusual prey. Over in Britain, we don't have any bird species small enough for any living dragonfly species to tackle.

Another alarming habit attributed to Common Green Darners, this one 100 per cent false, is that they can sew people's lips shut, and are in fact used by the Devil for that purpose. The 'needle' on the abdomen tip (actually the closely paired anal appendages) looks convincing enough for the sewing story to make an effective frightener for small children, but I am assured that grown-up Americans don't really believe this (well, most of them don't).

We were nearly 200 years too late to see one of the more improbable dragonflies on the British List – its only record was in 1818. The **Yellow-legged Dragonfly** or River Clubtail is a close relative of our own Common Clubtail, and hails from eastern Europe, with only a few outposts closer to us. It has declined and its geographic range has contracted eastwards since 1818, so it seems unlikely it will be back any time soon.

This is a shame, as the Yellow-legged Dragonfly is a most striking creature. Mostly lemon yellow, it has neat, clear-cut black markings, including a thickening double-stripe down the top of the abdomen, and a complex pattern on top of the thorax that looks like two scowling eyes if viewed head-on. The insect's actual eyes are vivid green, and its body is markedly less club-shaped than that of the Common Clubtail, and its legs are yellow at the tops, but black further down.

The other vagrant dragonflies on the British list all belong to the family Libellulidae, the chasers, skimmers and darters, and all have come to us from southern Europe. One of the most striking is the Banded Darter, known from one record in Wales in 1995. In fact it was one of two darter species recorded for the first time in August that year, which also saw influxes of other rare darter species.

The **Banded Darter** is a very distinctive species. It is small and the male has quite a pronounced waist. Like most darters the males are red-bodied, the females yellow-brown, and both sexes sport a broad dark band across all four wings. Its species name *pedemontanum* means 'foothills' — it prefers hilly habitats, which presumably explains why the individual in Wales picked an upland moor to end its days. Will there be more? It seems possible, given that the species has recently been found breeding in the Netherlands — prior to that its nearest breeding grounds were further south-east.

The other dragon to make its first appearance in 1995 was the **Scarlet Darter** or Scarlet Dragonfly. Unlike the Banded Darter, though, this one has been back a few times since, though not in 2011 or 2012. This dragonfly belongs to a different genus to the other darters, *Crocothemis*. This name is derived from the Greek word for saffron, and some of the other species in the genus are indeed bright yellow, but not this one. The male is the most arresting shade of intense, vivid scarlet-red imaginable from head to tail, much brighter than any of our other red dragonflies, and as such is completely unmistakable. The species name *erythraea* comes from the Greek word for blood — much more apt.

As well as its eye-popping colour, the Scarlet Darter is also bigger and fatter than other darters, closer in fact to a chaser in size and shape. The individual in 1995 was found at a pool on the Lizard, Cornwall, Britain's most southerly point. Cross over to mainland Europe and it becomes increasingly common as you head south, but a warming climate could see increasing numbers of this spectacular dragonfly showing up in Britain.

The year 1995 was a great one for **Yellow-winged Darters**. So was 1926, and 1945, and 1955. This dragonfly is subject to mass migrations from time to time, and in these years there were many records across the UK, but none of them resulted in established breeding populations, with just a few records in the years following the influxes. In this respect, the darter is rather similar to the Painted Lady butterfly, of which a jaw-dropping 1 billion were estimated to arrive from the south over a few short weeks in 2009. Painted Ladies don't seem to be able to survive the British winter at all in any of their life-stage forms. Yellow-winged Darters, though, do live and thrive in the much colder climate of Scandinavia, so why they can't seem to establish themselves here is a bit of a mystery.

The Yellow-winged Darter is quite a pale darter: the males bright orange and the females straw-coloured, both with an extensive yellow flush across the wing bases. Its species name *flaveolum* comes from the Latin word for 'yellow' — *flavus*.

The **Vagrant Darter** or Moustached Darter, *Sympetrum vulgatum*, has occurred in the UK a few times, including multiple records in the big year of 1995. However, it is so similar to the Common Darter that it may well be more frequent than we realise. I must admit that I didn't examine every single Common Darter I saw to check it didn't have a moustache. I have, however, just been through all my photos of the species, and there's not a tache to be seen.

The 'moustache' is actually a dark line that runs along the top of the frons and down the inner edge of both eyes, which would be an anatomically problematic moustache style for a human face. Maybe it's best not to think you're looking for a moustache but just remember that the frons is outlined on the left, right and top with black. The Ruddy Darter has a similar face pattern, but the Common just has black along the top of the frons. So ideally you would check each darter's face for this marking (no 'moustache' — it's a Common) and if there is a 'moustache' check the legs (all-black — it's a Ruddy). The combination of the tache and pale leg stripes adds up to a Vagrant Darter.

Its species name, *vulgatum*, implies it is a common species (from the Latin *vulgus* meaning 'common') which it is in central and north-east Europe. It has similar behaviour and habitat needs to the Common Darter, and as it lives in the same sort of climates we have in the UK — or colder — it might thrive over here if it ever was to become established.

There are British records of yet another rare darter, but this one bucked the trend by not showing up with all its relatives in 1995. In fact, the last British record of **Southern Darter** was back in 1901, and some of today's Odonata authorities consider that neither this nor the three previous records of the species are as kosher as they should be. It is a standard darter in appearance, the males light red, the females straw-coloured, and is best distinguished from the rest by a lack of black markings on the thorax sides. Southern Darters live around the Mediterranean, and their species name *meridionale* means 'southern'.

There is one dragonfly species that has more or less lost its place on the British List not through extinction but through reclassification. The '**Highland Darter**', a rather dark red (yellow in females) darter found in north-west Scotland and western Ireland is now generally regarded as not a full species in its own right but a local variant of the Common Darter. It has more extensive black markings on its thorax sides and on the underside of its abdomen, but there is a lot of variation from individual to individual.

As I saw for myself, the Common Blue Damselflies in the Scottish Highlands are distinctly blacker than their counterparts down south, and there are many other examples in the insect world of more northerly populations being blacker than southerly ones — probably an adaptation for maintaining a higher body temperature (black surfaces absorb the sun's warmth more readily). So it seems likely enough to me that the 'Highland Darter' isn't a separate species, but the best way to confirm this would be through DNA comparison tests, so we will probably have a definitive answer in the not-too-distant future.

The last of the dragons on the British list has the most splendid name — **Globe Skimmer**. As you would expect, this darter-like dragonfly is a noted long-distance traveller, and on a few occasions those travels have brought it to Britain, most recently in 1989. This is a medium-sized dragon, not unlike a female Scarce Chaser in appearance, even down to having little dark smudges on its wingtips. The sexes are pretty much alike, both light browny-orange with a dark central stripe on the abdomen. For its size it has mighty wings, proportionately very long, with very broad bases to the hindwings. Its scientific name *Pantala flavescens* translates loosely as 'yellowish thing that is all wings', a very apt description.

Globe Skimmers belong, by rights, in the tropics, on both sides of the Atlantic. They are more widespread than any other dragonfly on Earth, and also hold the record for highest-flying species, having been recorded blithely flying along at an astounding 6,200 metres. When on the move, the skimmers gather in huge swarms and use thermals (rising warm air currents) to gain height, and they will also make lengthy sea crossings. This willingness to travel makes the Globe Skimmer a real pioneer – for example, it is the only dragonfly to have colonised Easter Island. However, it is very rarely seen in Europe, and there is speculation that the prevailing wind across the Sahara makes a significant barrier for it. Therefore, the British records are considered by some to involve dragons that didn't fly here under their own steam but were imported in shipments of goods from further south.

Finally, there is one British dragon on the 'extinct' list, its story similar in some respects to that of the Norfolk Damselfly. This was the very lovely **Orange-spotted Emerald**, *Oxygastra curtisii*, once an established but very scarce resident in Dorset, where it frequented the Moors river, in the east of the county. There was also a record from the Tamar in Devon/Cornwall, but as only one individual was seen it's difficult to be sure if the species was actually established there. A pollution accident upstream of the Dorset colony in the 1960s seems to have led to its disappearance.

This dragonfly looks a lot like our other emeralds but its scientific name indicates a more distant relationship. It is the only species in the genus *Oxygastra*, which as far as I can make out means 'sharp stomach'. Its species name honours John Curtis, an English entomologist of the 19th century who wrote and illustrated *British Entomology – being illustrations and descriptions of the genera of insects found in Great Britain and Ireland*. This was a work of great importance, and a beautiful one too. The Orange-spotted Emerald has its place in the book, with a lovely illustration of an adult in flight alongside a water-rush of some kind.

So those are the ones that got away — some were almost in my reach, others nowhere near. Those near misses — the likes of Southern and Scarce Blue-tailed Damselfly — do feel like unfinished business. But the unpredictability of the vagrants' arrival means that you would need ludicrous amounts of luck to see all the species on the British List if you devoted a whole lifetime to it. Unless, of course, you go abroad to see them ... and no doubt that would be the start of a whole new adventure, for the species on the British List make up only about 1 per cent of the total number of Odonata species alive on Earth today. Anyone for a Jaunty Dropwing, an Ivory Pintail or a Royal River Cruiser? Then I'll see you on the other side of the world!

Epilogue

Before I began my dragon-quest, I underestimated how much influence the weather would have on our hunt. As a birdwatcher first and foremost, I didn't give enough consideration to how much insects are at the mercy of weather, and the two dragon-seasons of 2011 and 2012 had plenty of problematic weather. The hot and early spring of 2011 brought out some species weeks earlier than expected. Then the weeks of chilly, rainy weather that served as 'summer' in 2012 made it hugely challenging for insects to survive and breed — and for insect-watchers to see them.

An adult dragonfly or damselfly needs the air temperature to hit a certain level before it is capable of strong flight. Its prey species — smaller flying insects — also need warmer air before they will be active and catchable. So dragons and damsels won't hunt if it's too cold. Nor will they be active when there are strong winds, or heavy rain. If the weather is bad enough prevent activity, they will simply sit it out in a sheltered spot, expending almost no energy, and can survive days and days of inclement weather like this, although the longer they have to sit and wait, the higher their chances of being taken by an opportunistic — and warm-blooded — predator. As long as this doesn't happen, a dormant dragon will spring back to life and resume its hunting activity very quickly when the sun comes out.

When things really warm up, dragons and damsels have no trouble hunting, and can engage in life's more energy-expensive activities — defending a territory, securing a partner, mating and egg-laying. Extremely hot weather may curtail activity, and long hot summers carry the risk of drying up the dragons' nurseries, the streams and ponds in which they lay their eggs. But within a typical dragon or damsel's adult lifespan in a typical British spring or summer, there will be enough hours of good-enough weather to allow completion of the reproductive cycle. When it comes to survival of the immature stages through winter, as long as the water bodies they live in don't freeze up completely they should be fine, and in a typical

British winter even smaller ponds will not freeze completely in a cold snap — there's still plenty of oxygen-rich fresh water under a crust of ice. The trouble, though, is that 'typical' British weather seems to be changing.

As I write, it is late March 2013 and spring should be well underway, but much of the country is still buried under a snowy blanket and temperatures are sub-zero. An article I read online this morning told me that last year's rainy summer had a catastrophic effect on nearly all British butterfly species. Another news story on the same website said that we had our hottest October day ever in 2011. It seems that wildly variable and unpredictable weather patterns may be the 'new normal', as climate change takes hold. Hard times for humans, and wild animals will simply have to adapt or die. What will this mean for our dragons and damsels in years to come?

We are already seeing increasing numbers of species from southern Europe or even further south. Spring 2011 saw a major influx of Vagrant Emperors, and late summer 2012, Southern Migrant Hawkers. But while moving north in a warming world might help southern pioneers find places where temperatures suit them, ecosystems, different day-length patterns and habitats may not be right at all. And what about species like Northern Emerald and Azure Hawker, restricted to northern Scotland and adapted to colder conditions? They have no geographical space to go north. Species with very short flight seasons, such as Norfolk Hawker, are also vulnerable, as they have less time to get lucky with the weather.

The species that are already doing well will probably continue to do so. These have a wide distribution, a long flight season that gives them opportunities to disperse, and an ability to use all kinds of different habitats. Species like Four-spotted Chaser, Blue-tailed Damselfly, Common Blue Damselfly and Common Darter seem likely to be with us in abundance for the near future at least. They may be joined by newcomers from the south — Migrant Hawkers and Small Red-eyed Damselflies seem to be progressing northwards through Britain with particular vigour. However, northern species and those with limited habitat choices are at risk.

APPENDIX I: TECHNICAL DETAILS

Without my binoculars, I'd have had a hard time finding and identifying some of the dragonflies encountered. As a keen birder, I had a pair of 10x42 Leica Trinovids, though if I had been on the hunt for some that would be perfect for dragonfly-watching I would instead have opted for a model that offered really good close focusing.

Taking photographs of the dragons and damsels that we found was a major objective. For a couple of years, Rob had a very versatile Panasonic bridge camera with a large zoom range and a macro mode, meaning that in theory it was ideal for wildlife photography, able to handle distant birds and tiny insects with equal aplomb. As his interest in photography grew, Rob bought a DSLR, a Nikon D300 with 18–200mm lens. I inherited the Panasonic, and used it constantly for more than a year before I too felt the need to explore the possibilities of DSLR photography so I bought an almost unused D300 body from a friend.

I bought one lens for this camera – a 300mm prime. I often use it with a 1.4x teleconverter, which gave me 420mm – a magnification not far short of standard birdwatching binoculars. This suited me as bird photography was my main interest, but the set-up has also proved very good for Odonata, especially the larger dragonflies which are restless, flighty and best photographed from a fair distance. The next was a 180mm macro lens which allows you to focus at very close range and take life-size photos of insects, still at a good distance away – great for damselflies. Rob also has a 90mm macro, which provides 1:1 close focusing, although because it is a smaller lens you need to be much closer to the subject to achieve this.

With this kit we were well-equipped to photograph Odonata in most circumstances. If a dragon was resting on the far side of a river and we couldn't get close, the 300mm (plus converter) would be the lens of choice. Without the converter, this lens focuses pretty quickly and this allowed us to take a few flight shots. For more tractable resting insects the 180mm or 90mm was best, and they allowed for some pretty extreme close-ups on the few occasions we were lucky enough to find a big dragonfly that was docile enough for a very close approach. Rob experimented with flash photography now and then with a very obliging subject and used a Nissin flashgun, though natural light was more than adequate in most cases.

Other, less essential kit included my rucksack with incorporated folding chair, allowing for low-level photography any time, anywhere. For kneeling or even lying down to take eye-level photos when the subject was sitting on the ground, kneepads and/or a roll-up foam mat would have made either of these more comfortable.

APPENDIX 2: WATCHING AND
PHOTOGRAPHING DRAGONS AND DAMSELS

Watching dragonflies and damselflies is pure enjoyment – unless you have a phobia of fast-flying insects, in which case you're probably reading the wrong book. Taking photographs of them, though, can be an exercise in total frustration. If you really do like to challenge yourself then taking up dragonfly and damselfly photography will guarantee you a lifetime of it – and in some circumstances taking good Odonata photos is not really difficult at all.

To simply enjoy watching dragons and damsels, pick any warmish day, with sunshine and not too much wind, and visit a suitable bit of habitat. An air temperature of at least 15°C is needed for most Odonata to be active, though large dragons can cope with slightly cooler temperatures, and damsels prefer it a little warmer. In most species, territorial and mating behaviour happens towards the middle of the day, so this is when you'll see most action at the waterside. Damselflies tend to go to roost fairly early in the evening, at least two hours before sunset, but some dragonflies will continue to hunt up until dusk or later. For a chance of seeing the process of metamorphosis as a nymph transforms into a dragonfly, try searching along the margins of lakes and rivers in the morning, carefully checking upright stems of emergent vegetation.

Binoculars will allow you to track a flying dragon for longer and further than you could without them, and will reveal details that may be lost with a more distant view. Choose binoculars that allow you to focus at close distance, and with a good depth of field so you don't have to keep adjusting the focus as the dragonfly or damselfly goes further away or comes closer.

The best tools you can have for good Odonata photographs are patience, a bit of creative flair, and some understanding of your subjects and how they live. Rob and I had no particular expertise at photographing Odonata at the start of 2011, The DSLR camera gear we have is in the 'pro-sumer' bracket, aimed at keen amateurs and skint professionals, but it is certainly possible to take excellent dragonfly and damselfly photos with ordinary compact cameras. Compacts often have very good macro modes, so you can take detailed images of small subjects, provided they sit still.

Any warm, sunny and reasonably still day from April through to October will encourage Odonata activity, and a walk anywhere around fresh water should provide plenty of options for photography. Ideally you'll be able to get right down to the water's edge, as that's where some of the action will be. The closer to the middle of the day it is, the higher the chances of seeing lots of activity, as many species spend just a few hours a day courting and mating.

In the early morning, dragonflies and damselflies are usually quite inactive, as they wait for the air temperature to rise. Brand new dragonflies of the larger species tend to make their emergence from the water at night so may be found in the morning still sitting on their exuvial cases, expanding their wings prior to their maiden flights. Fully developed adults will have found roosting places the night before, and stay put until they have properly warmed up. So early morning can be the perfect time for static photography – but actually finding your subject may present a challenge. Hawker dragonflies often rest high up in trees, and even if they are lower down they can be hard to spot as when their bodies are cool their colours are less brilliant, and they are adept at choosing spots where they are surrounded by colours similar to their own.

If you go dragon hunting early in the morning you should move around slowly and carefully, using binoculars to scan ahead. Get too close and/or move too suddenly and even a sleepy dragon may be startled into flight – though if it is very chilled it will allow you very close and you might even be able to pick it up. But if you do, take its photo first, in case it objects.

Damselflies often go to roost perched on a vertical stem of some kind, one that's thick enough to hide their bodies, but slim enough for them to be able to see around it with their very widely spaced eyes. Thus they have the advantages of concealment and all-round visibility. Try searching among stands of rushes at the waterside, or in long grass in fields close to water.

Finding dragonflies and damselflies in the warmer hours of the day is easier. All damselflies and most dragonflies spend a fair amount of their active hours perched, looking out for prey or possible mating partners. You can get close enough for photos, provided you are careful and slow in your movements. I've noticed in nearly all species that for a moment after landing, they remain in an alert posture, ready to immediately take off again if for whatever reason they decide it's best not to stay at that spot. Then their posture changes a little, the legs bend and the abdomen dips down as they relax. Once your target has 'dipped' into this more relaxed posture, you can start closing in.

The insect can see you coming, but you have to convince your target that your movement doesn't represent a threat. So close in slowly and directly, keep your camera raised and arms tucked in and try not to go across the insect's sight line, just forwards within it. With each increment of increasing closeness, take a photo, even if you plan to get closer.

If you disturb a damselfly or a darter dragonfly, it will probably soon land again nearby so you can try again. Chaser and skimmer dragonflies will very often return to the exact same spot, if you give them a little space to do so. However, a hawker will probably fly off at high speed, never to be seen again. So if you do find a perched hawker, be particularly careful when trying to get close to it.

Once you have found the dragonfly or damselfly you want to photograph you can just point and shoot immediately. This is what I do, to ensure I actually do get at least one photo. If the insect is still there, then I'll try to take more, and better photos. First ensure that the photo will be in focus. If there's good light, you will probably already be achieving fast enough shutter speeds for sharp photos. If the shutter speed is too slow, then movement can blur the image. A good guideline is that the speed in 100ths of a second should be double the focal length of your lens in millimetres. So if I'm using my 300mm lens, I want shutter speeds of at least 1/600th of a second.

I'll also do what I can to stabilise the camera, and myself, to keep camera shake to a minimum. The simplest adjustments can make a big difference. Plant your feet solidly, shoulder-width apart. Support the camera with both hands, one on the body and the other on the lens. Tuck your elbows in. Lean on a tree, or a friend. Put the lens on a branch or railing. Kneel down and rest your elbows on your hips. It all helps.

A lot of insect photographers use tripods to stabilise their cameras. This gets rid of camera shake, and so allows them to use slower shutter speeds than if they were hand-holding the camera. The value of doing this is that they can 'stop down' – use a smaller aperture. When using an aperture of, say, f8 rather than f4, a photo will need a longer exposure, but it will have more depth of field, which means it's easier to get an entire dragonfly or damselfly in focus. The closer you are to your subject, the more depth of field you need to achieve this. Even when I try to position myself perfectly side-on to my subject, I often find that my more close-up photos show the insect's head in lovely sharp focus, but its tail tip is blurry. A bit more depth of field would sort out that problem. And the tripod not only keeps everything stable but allows you to take your time and be very precise when composing your photos.

The disadvantage of using a tripod is that by the time you have got it all set up, your dragonfly or damselfly may well have flown away. Or you may find a much better photo opportunity nearby but be unable to take advantage of it because all your stuff is set up in slightly the wrong place. For these reasons, I never use a tripod. Rob, however, does from time to time and has taken some great shots this way, when the subject has been very co-operative. More portable alternatives to a tripod include the monopod – like a tripod but with only one leg, so don't let go of it, but it's still connected to the ground so is inherently more stable than hand-holding. There's also the option of a beanbag, which you can put down on any surface and it will provide a safe and balanced support for a heavy lens.

If you find are able to get very up close and personal with an obliging Odonata, you can think about what kinds of photos you'd like to take. While the conventional side-on or top-down shot of the whole insect is always going to be most popular, there are lots of wild and wacky alternatives. It is always good to get eye-level with your subject, for a more intimate feel to the shot. Head-on photos are fun, and a very relaxed insect may make them even better by giving its eyes a rub-down with its front legs. The tracery of veins in the wings makes for pretty photographs, and a 'from below' perspective transforms an innocuous small damselfly into a looming monster.

If your subject will let you, it's worth experimenting with different angles by moving around to the other side, as the light direction can really change the whole feel of your photo. The 'best' lighting is with the sun behind you, so the subject is fully lit – for the clearest, brightest pictures with the most realistic colours – that's the angle to go for. But backlit and side-lit photos can have a very attractive arty look to them, even if the fine detail of the subject is lost.

You don't have to get right up in your subject's face to take a pleasing photograph, either. Showing the insect as the small object that it is, with elements of its natural habitat all around it, can work beautifully. Some of my favourite photos show the dragon or damsel quite small in the frame of the photograph, but this kind of photo only works well if everything else in the frame looks as good as the insect itself. So look out for dragons or damsels that are perched on interesting-looking things. Damselflies will often oblige by sitting on a flower or prettily dangling grass head, and if you don't like a damsel's chosen perch then you can gently encourage it to move on – it will probably settle again nearby and with any luck will choose a more photogenic spot. Darters like to settle on wooden objects as these tend to get rather warm, so look out for natural or artificial bits of bare wood with pleasing colours or textures.

We always hoped to take some 'action shots' of our subjects, ranging from static behaviour like munching on prey, to the trickiest of all, the flight shot. I'd be the first to admit that I need a lot more practice with the latter, but getting photographs of other interesting Odonata behaviour can be quite straightforward.

When it catches its insect prey, a dragon or damsel will usually land to eat it, though large dragons that have caught something small will gobble it up in flight, only landing if the prey is too big and wriggly to be dealt with on the wing. If you are watching a dragon or damsel and it makes a short flight before landing again, chances are it has caught something. Taking photographs of the ensuing carnage may not appeal to everyone, but it does give the opportunity to capture the remarkable Odonata eating apparatus in action. You'll still need to be careful when getting close though, as the insect will be just as alert and wary as when it is empty-handed. Usually the prey is a lot smaller than the predator, but sometimes both dragons and damsels will take on insects that are close to their own size and you may then witness a prolonged battle – particularly if it is a dragonfly versus dragonfly situation.

Of course, dragons and damsels themselves fall prey to other, larger or more cunning predators, and this can make an interesting photographic subject. Damselflies often end up in spider's webs and a damsel that has lost its struggle in a spider's web can make a good picture in death.

Some spiders don't make webs but instead are ambush hunters, either hiding or using expert camouflage to remain unseen until a hapless insect gets within grabbing distance. Crab spiders are camouflaged to match the flowers in which they lurk – they come in white, pink and yellow. Their main prey are the insects that come to flowers to feed on nectar – hoverflies, bees, moths and the like – but if a damselfly paused for a rest on their flower that would do just as well. Look out for crab spiders on large, open flower types – dog roses, ox-eye daisies and the like.

Birds are important predators of Odonata. Getting photographs of a bird actually taking a dragon or damsel is usually far from easy. In spring and summer, parent birds are collecting food for their chicks and so may be seen carrying squashed insects in their bills. One day at Dungeness, I photographed a female Reed Bunting with at least four damselflies packed into her bill – and this was on a day when I had seen hardly any Odonata at all.

Small birds usually aren't quick or strong enough to take down a big dragonfly, but birds of prey manage better. One in particular, the Hobby, is an expert dragon-catcher, and Hobbies tend to gather at places where dragonflies are abundant. In late summer and early autumn, wetlands in southern England that abound with Migrant Hawkers may attract lots of hungry Hobbies. If you have a long telephoto lens, quick reflexes, lots of patience and a bit of luck, you can capture the Hobby in action, bringing its feet forwards and then quickly swiping them back to catch a dragon, and then deftly dismantling the poor insect by nipping off its wings and sending them spiralling earthwards, while the wingless body gets eaten.

Most of our Odonata don't do very much in the way of courtship behaviour. It usually goes like this – boy sees girl, boy immediately flies over to girl and tries to grab hold of her and lock his back end onto her head, girl either allows this to happen or flies away as fast as she can. However, some do demonstrate a little light flirting, and males of many species have territorial encounters with other males.

The demoiselles, with their highly coloured wings, show off their best feature to each other with a rapid open-and-shut flick, repeated regularly. This is tricky to capture on camera, and when I tried to do it I was grateful for a) a large memory card in my camera and b) six frames per second – I just started taking photos when I thought the demoiselle was about to flick, rather than try to catch the blink-of-an-eye moment when it did. When a demoiselle is busy wing-flicking, pay attention to the intervals between flicks to predict when the next one will happen. Going for a head-on angle is good for this behaviour as it shows off the spread of wings to their best advantage.

The mating behaviour of Odonata is easy to see and pretty easy to photograph. A lakeside in June on a sunny lunchtime will probably be alive with hormonally fired-up damselflies, all pairing up, trying to pair up or trying to muscle in on others that are already paired up. Once a male has got hold of a female and locked his claspers onto the back of her head, the two fly in tandem to a place where they can sit in peace and get on with mating. So if you spot a pair in tandem, follow them to where they land and you should be able to get some photos of the mating process. Be careful of composition – with wings and abdomens pointing in all directions, it's easy to accidentally crop out something important and you may need to stand further back than you would when photographing a single damselfly.

Unpaired males tend to be very interested in happy couples, sometimes attempting to forcibly wrestle the poor female away from her original partner, so look out for this behaviour. Coupled-up pairs are usually slightly more approachable than singletons, being distracted by each other and also being more reluctant to fly, but they will still scarper if you aren't careful.

Dragonflies are a bit more discreet about mating than damsels, and in most species the encounter is also less prolonged – sometimes it's just a few seconds of mid-air bliss before the couple part forever, but in many species a joined pair will land and do the business for several minutes or longer. If you find a male dragonfly in patrol mode, flying restlessly up and down the same stretch of waterside, keep an eye on him, as he will probably spot an approaching female before you do.

Once a successful mating has occurred, the female is ready to lay her eggs, which usually involves dipping her rear end into a suitable bit of water. This is another interesting quirk of behaviour you might want to photograph, and is one of the few where the hawker dragonflies are actually a bit easier than most to catch in the act. A female hawker usually perches to lay her eggs, and stays sitting for a while on some floating vegetation with her abdomen submerged. Check in places where there is some overhanging vegetation, as females like shelter and privacy when engaged in this activity.

Most damselfly females lay their eggs while the male who fertilised them is still attached. So she drops down to water level, settles on something floating and dips her back end in the water. While she is doing this, the male pokes up vertically from her neck, with his legs clasped together around nothing, looking like a bizarre bodily appendage. Sometimes dozens of joined-up pairs will use the same small area, which can make for great photographs.

In some other species, such as the demoiselles and emerald damselflies, the males do detach themselves when the female is ready to lay her eggs, but they stay at her side, 'guarding' her jealously from other males' attention. This mate-guarding behaviour is an interesting aspect of Odonata life, but can be difficult to photograph as the two insects are probably going to be on different focal planes from most of the angles you could take your pictures. If this is the case, you'll need a lot of depth of field to get sharp photos – this means stopping down the lens, which means longer shutter speeds, which in turn means you'll need good light and/or a tripod to keep the camera steady. On the upside, the pair of demoiselles or emeralds will often sit for a long spell in one spot as the female lays her eggs, which will give you enough time to mess about with angles and camera settings.

The darters egg-lay while still joined in tandem, but rather than sitting down the female just bobs down towards the water and dips her backside in briefly, without stopping. The pair will work their way across a lake in this manner, dipping down every few seconds. Common Clubtails, emeralds, chasers and skimmers and also Golden-ringed Dragonflies lay their eggs without the male attached (though he is probably loitering nearby to keep an eye on the fate of his genetic material) but they do it in flight as well. The female flies low over the water and dips her bottom quickly in the water. In the case of the Golden-ringed, the movement is surprisingly forceful, more of a stab than a dip. As this is such an active and unpredictable process it is quite tricky to photograph, although perhaps not as challenging as capturing a dragon in full-speed flight. You may have more success if you back off a little bit so there is more room in the image frame to track the moving insect/s.

Catching a dragonfly in flight is difficult, and here is where having a DSLR camera comes into its own, as DSLRs don't have the 'shutter lag' which can cause the crucial moment to be missed, and their lenses are generally very quick to focus. But even with the best gear it's still close to impossible to catch a flying dragonfly that is weaving about in unpredictable directions, especially if it's flying against a complicated backdrop – say, riverside vegetation – which will confuse the camera's autofocus system.

The 'easiest' dragonflies to photograph in flight are those that periodically pause and hover on the spot. One example is the Downy Emerald – males of this species track around a lake shore and stop now and then to hover more or less motionless for three or four seconds. That's still not very long, and it took me the best part of half an hour to get my first flight photos of the species. My most successful shots were taken with autofocus, but I did have a go with manual focus as well. Maybe if I had practised more with manual focus before I met this dragon I would have had more success, but my eyes just weren't as accurate as the camera in picking the point of sharpest focus.

Some dragons aren't so much given to hovering, but do patrol a regular beat, and if you wait you will have multiple fly-bys and opportunities to photograph the dragon as it goes by. If you are alert you'll see the dragon coming from a long way off and be ready to catch it when it is closer. A longish telephoto lens: 300mm or 400mm, will serve better than a macro lens, as most macros don't focus quickly enough.

I would advise using autofocus, and pre-focusing your lens on a spot close to where you think the dragonfly is going to be, so that when it comes into your viewfinder it is not too far off being in focus – this will help the camera's autofocus system to lock onto it more quickly. I would also recommend standing in a position where you'll be looking at the dragonfly against the sky or water, rather than against vegetation – a relatively clear, featureless background will help the autofocus find its target more easily. I'd also recommend not hurling your camera into the water in frustration when you've spent the best part of the day trying and failing to take even one sharp flight shot – the struggle is character-building, and you will get there in the end.

GLOSSARY

ABDOMEN The third and largest of a dragonfly or damselfly's three body parts, usually elongated. It is formed from 10 segments and contains the gut and sexual organs.

ANAL APPENDAGES Two paired appendages at the abdomen tip, which are usually shaped like little callipers. In some species they are very small, in others quite large and prominent. The male uses his to grip the female's head during mating.

ANTEHUMERAL STRIPES Longitudinal markings on the top of the thorax, seen in many Odonata species – their shape can be an important aid to identification.

CLASPERS Another word for the anal appendages of a male Odonata, which he uses to grip the female's head during mating.

COSTA The thickened vein along the leading edge of a dragon or damsel's wing.

CUTICLE The tough outer covering of an insect's body.

EXUVIA The empty husk of a dragon or damsel nymph, left behind when the adult insect emerges.

FAMILY A grouping of closely related genera. Family names are written in plain text, not italics. For example, the damselfly family Coenagrionidae contains several genera including *Coenagrion*, *Ceriagrion* and *Ischnura*.

FRONS The 'face' of a dragon or damsel, a flattened squarish panel between the eyes and above the mouthparts, most evident in a head-on view.

GENUS (PLURAL GENERA) A grouping of closely related species. Identified by a scientific name (written in italics) – for example, the damselfly genus *Coenagrion*.

IRIDESCENCE Bright reflected colour with a metallic appearance, caused not by a coloured pigment by the structure of the cells in an insect's cuticle that reflect certain light wavelengths when viewed from the right angle. Iridescent dragons and damsels look all dark from some angles.

MACRO LENS A camera lens designed to take very close-up photos of small objects.

MOUTHPARTS The eating apparatus of a dragon or damsel.

NYMPH The immature stage of a dragonfly or damselfly, which does not fly but lives underwater. Also called a larva.

ORDER A grouping of closely related families. The order Odonata includes all dragonflies and damselflies, which each represent suborders within Odonata.

OVIPOSITOR The egg-laying tube seen in some female dragons and damsels. In some species (for example the Golden-ringed Dragonfly) the ovipositor is long and sharp, and is used to cut into underwater vegetation before eggs are laid.

PIGMENT Substance in an insect's body that provides colour, by absorbing some light wavelengths and reflecting others.

POLYMORPHIC Occurring in several distinct forms or morphs. In some damselfly species the females have several colour morphs.

PREDATOR Any animal that catches and eats other animals.

PREY Any living animal that is caught and eaten by a predatory animal.

PRUINESCENCE A pale blue, dusty coating or 'bloom' that forms on the bodies of some mature male dragonflies and damselflies.

PTEROSTIGMA A thickened, coloured cell on the outside, leading edge of the wing, which aids stability in flight. The colour and shape of the pterostigma can be an important aid to identification.

SCIENTIFIC NAME The two-part name of a species which is universal in all languages – the first word is the genus name, the second the species name. It is always written in italics. Sometimes called the 'Latin name' but this is inaccurate as often the words are derived from Greek or other languages.

SETAE Fine hairs that cover a dragonfly's or damselfly's body.

TAIL The tip of the abdomen. Not a rigorous or precise term, as insects don't have 'tails' in the same sense that, say, mammals do. The 'tail' of a dragonfly may refer to one, two or more of the abdomen segments and is usually used when the abdomen tip is a contrasting colour to the rest – for example in Black-tailed Skimmer.

TANDEM The position of a joined pair of dragons or damsels before or after they mate, when the male is gripping the female's head with his anal appendages.

TARSUS The outermost section of a dragon or damsel's leg, including the claw at the tip.

TENERAL A recently emerged dragonfly or damselfly, which has yet to attain its mature coloration.

TERRITORY A patch of habitat that an animal defends against intruders (usually, but not always, of its own species).

THORAX The middle of an insect's three body segments, between the head and abdomen, to which the wings and legs are attached.

TIBIA The middle section of a dragon or damsel's leg.

WHEEL The circular or heart-shaped mating position of a pair of dragons or damsels, the male clasping the female's head and the female bringing her abdomen forward to make contact with the male's secondary sexual organs.

RECOMMENDED READING

Field Guide to the Dragonflies and Damselflies of Great Britain and Ireland by Steve Brooks and Richard Lewington (1997). British Wildlife Publishing. Gillingham, Dorset

This is *the* Odonata field guide for Britain, with tremendous detail and illustrations that are second to none. It is an exemplary field guide by any standards, with very detailed descriptions covering behaviour, biology and conservation as well as identification. There is also a very useful gazetteer of sites.

Watching British Dragonflies by Steve Dudley, Caroline Dudley and Andrew Mackay (2007). Subbuteo Natural History Books. Shrewsbury.

A very useful book that combines field guide and site gazetteer. includes field guide pages but places more emphasis on (and devotes more pages to) descriptions of the best dragon and damsel sites and how to get the most out of them.

Field Guide to the Dragonflies of Britain and Europe by Klaas-Douwe B Dijkstra and Richard Lewington (2006). British Wildlife Publishing. Gillingham, Dorset.

If and when your dragonfly exploits take you beyond British shores, make sure you have this guide with you.

Britain's Dragonflies – a field guide to the damselflies and dragonflies of Britain and Ireland by Dave Smallshire and Andy Swash (2009). WildGuides. Basingstoke.

If you prefer your field guides to be illustrated with photographs, this is the book for you. The book includes several 'potential vagrants' – European species which are spreading north and are poised to make their first appearance in Britain.

British Dragonfly Society website: www.british-dragonflies.org.uk. A tremendous wealth of material is available here, from detailed photographic guides to species identification to up-to-date reports of sightings.

Wild About Britain website: www. wildaboutbritain.co.uk. An online community for everyone interested in wildlife, with forums, a photo gallery, blogs, gear reviews and more.

I also gleaned some very useful advice on locations and timings from various wildlife blogs.

ACKNOWLEDGEMENTS

I'd like to thank all of the team at Bloomsbury for backing this project, especially Lisa Thomas whose support, enthusiasm and kind tolerance of the various wobbles on the way was beyond priceless. Big thanks to Nicola Liddiard for creating the lovely design, to Paul Liddiard, who proofread the final version and Marie Lorimer for the index.

My heartfelt thanks goes out to all my fellow Odonata-philes online, who made life so much easier by publishing details of their sightings in blogs, on message boards and on the sightings pages of the British Dragonfly Society's website. Also thanks to all the people I met 'in the field', whose knowledge helped me to find the species I was looking for, and in some cases helped me ascertain that the dragon or damsel I was looking at was (or just importantly was not) what I thought it was.

Many friends provided moral and practical support to me and to Rob while we went on our dragonfly and damselfly hunts, and some even came with us. Thanks to Steve and Pauline for looking after Mittens when needed. For their company on various dragon-hunts, thanks to Mike, Kathy, Flo, Graham, Shane, Phil, Michèle, Carol, Susan, Nigel and Jim. A special big thank you for Susan and Michèle for all the hand-holding and invaluable practical help when things went a bit pear-shaped in the summer of 2012. Thanks to my family for being great, and to my aikido pals and online friends at the RSPB forums and the LLL forum for encouragement and especially for all of you who shared your Odonata anecdotes with me.

Finally, thanks to Rob for all of the support, driving, company, photographs and general team-mateyness over the last two years – I really couldn't have done it without you.

LIST OF LINE ILLUSTRATIONS

INDEX